SMALL-SCALE SYNTHESIS
of LABORATORY REAGENTS

with Reaction Modeling

SMALL-SCALE SYNTHESIS
of LABORATORY REAGENTS
with Reaction Modeling

LEONID LERNER

CRC Press
Taylor & Francis Group
Boca Raton London New York

CRC Press is an imprint of the
Taylor & Francis Group, an **informa** business

CRC Press
Taylor & Francis Group
6000 Broken Sound Parkway NW, Suite 300
Boca Raton, FL 33487-2742

First issued in paperback 2019

ISBN-13: 978-1-4398-1312-6 (hbk)
ISBN-13: 978-0-367-38304-6 (pbk)

Library of Congress Cataloging-in-Publication Data

Lerner, Leonid.
   Small-scale synthesis of laboratory reagents with reaction modeling / Leonid Lerner.
      p. cm.
   Summary: "The in-lab preparation of some chemical reagents provides a number of advantages over purchasing commercially-prepared samples. This volume contains a detailed description of methods for the rapid and reliable synthesis of many useful reagents that are otherwise difficult to obtain. It provides spectroscopic analyses of products, presents the thermodynamic/kinetic data of reactions, and offers a thorough analysis of the purity of the final products. Enhanced by useful diagrams and photographs, this text is a valuable reference for researchers in small or remote laboratories"-- Provided by publisher.
   Summary: "When working in a small chemical laboratory attached to a non-chemistry specific project one is frequently asked to solve problems which occur in the main project. Usually one can think of several solutions and experimental testing is required to choose between the different approaches, or indeed to find an approach that works. These often require the use of small amounts of chemicals which the laboratory does not stock. Obtaining these chemicals from commercial suppliers in a suitable timeframe sometimes poses problems. This is particularly the case in isolated regions such as Australasia, when the supplier has no local stocks of hazardous substances and they have to be brought in from overseas, something which can introduce delays from several weeks to six months or more. Since there is no certainty that the substance whose acquisition is introducing the delay will form part of the ultimate solution, this delay is particularly objectionable. The primary aim of this book is therefore to provide reliable laboratory syntheses of the most common reagents whose acquisition can introduce delays due to their hazardous nature. It is apparent that there is little point in replacing delays in reagent purchase with uncertainties in their synthesis, hence the preparations presented in this book were chosen with the following criteria in mind: short duration, availability of apparatus, high yield, and high purity of the product. It was also considered necessary for the syntheses to be described in greater detail than is normally provided in the literature, since I have frequently found that omission of apparently small details can subsequently waste much time"-- Provided by publisher.
   Includes bibliographical references and index.
   ISBN 978-1-4398-1312-6 (hardback)
   1. Chemical tests and reagents--Synthesis. I. Title.

QD77.L47 2011
543.028'4--dc22

                                                         2010038460

Visit the Taylor & Francis Web site at
http://www.taylorandfrancis.com

and the CRC Press Web site at
http://www.crcpress.com

Contents

Preface

When working in a small chemical laboratory attached to a nonchemistry-specific project, one is frequently asked to solve problems that are occurring with the main project. Usually, one can think of several solutions, and experimental testing is required to choose between the different approaches, or indeed to find an approach that works. These often require the use of small amounts of chemicals that the laboratory does not stock. Obtaining these chemicals from commercial suppliers in a suitable timeframe sometimes poses problems. This is particularly the case in isolated regions such as Australasia, when the supplier has no local stocks of hazardous substances and they have to be brought in from overseas, something which can introduce delays from several weeks to six months or more. Since there is no certainty that the substance whose acquisition is creating a delay will form part of the ultimate solution, this delay is particularly objectionable. The primary aim of this book, therefore, is to provide reliable laboratory syntheses of the most common reagents whose acquisition can introduce delays due to their hazardous nature.

Since there is little point in replacing delays in reagent purchase with uncertainties in their synthesis, the preparations presented in this book were chosen with the following criteria in mind: short duration, availability of apparatus, high yield, and high purity of the product. It was also considered necessary for the syntheses to be described in greater detail than is normally provided in the literature, because I have frequently found that omission of apparently small details can subsequently waste much time. These requirements are reflected in the experimental sections of this book as follows:

- All syntheses can be accomplished in an afternoon, with sourcing and preparation of apparatus requiring an additional day or part thereof.
- Yields are generally in the range 80 to 90%. The few lower-yielding preparations (e.g., 45% for cesium in Chapter 5) are included when lower yield does not influence reagent purity and is offset by the cost differential with respect to the purchased product.
- The purity of the reagents is generally >98%, and an assay of most reagents, including the IC, NMR, and GC/MS spectra, is presented in the Appendix at the end of the book.
- The apparatus is available in most modestly equipped laboratories. The few more specialized items are the quartz tube for the high-temperature syntheses (Chapters 5, 6, and 20), and the MgO crucible of Chapter 3.
- An unusually high level of detail is provided, with all experiments having been performed by the author.

The second aspect of this book is educational. It deals with fundamental processes in inorganic chemistry, and while most syntheses are based on reactions previously investigated, or at least reported, the preparations, and a good part of the discussions are original, and not a reformulation of previously available material. This has been possible for several reasons. Firstly, many syntheses are based on processes cursorily described in old literature, often without accurate determination of experimental conditions and yield, let alone an analysis of product purity or an investigation of the influence of variation of experimental conditions on the progress of the experiment. Some of the syntheses presented in this book make use of computer control, which enables experimental conditions to be tightly monitored. For instance, the natural tendency for thermal runaway of a molten potassium hydroxide cell in Chapter 3 is controlled by a negative feedback algorithm applied to cell current.

Secondly, the relevant literature often affords even less consideration to theoretical aspects of the experiment. While books on synthesis do not generally delve into theoretical considerations, or contain much in the way of calculations, I have endeavored to develop sufficient analysis relevant to an experiment when this has important consequences for synthesis parameters, such as the reaction time, yield, and purity of the product. Thus, the reader will find answers to questions not previously covered in the literature, for example:

- Why the electrolysis of water in a sodium hydroxide (Castner) cell, which consumes half the cell current and is a major drawback of the process commercially, is essential to cell operation.
- Why differences in the physical properties of sodium and potassium make the production of potassium in a Castner-type cell impossible.
- How sodium sulfate can be used to break the 98% H_2SO_4/H_2O azeotrope, and the dependence of oleum concentration and yield on temperature.
- How the photolytic chlorination rate in liquid and gas phases of carbon tetrachloride depends on concentration and ultraviolet (UV) intensity, and the effect on the reaction rate in the liquid and gas phases.

It should be stated that the discussion of the theoretical aspects of the syntheses is in no way extensive or complete. Due to the large number of preparations such a task would not be possible within the confines of the present book. The extent to which a reaction is analyzed is governed mainly by the degree to which it was deemed necessary to increase the performance of a given synthesis, and where a unified description relevant to the synthesis has not been found in one place in the literature.

The present book assumes knowledge of chemistry, and some aspects of physics and mathematics to at least introductory university level, while those performing the syntheses should have practical chemistry training to at least advanced undergraduate level. The book should be of interest to anyone engaged in practical synthetic chemistry as well as those interested in fundamental reactions in inorganic chemistry. It can also be used as a project source for advanced courses at universities and institutes of higher learning.

I wish to thank Professor Kevin Wainwright and Flinders University for cordial hospitality and provision of resources to enable this book to be completed, as well as the Defence Science and Technology Organisation. I also wish to thank my wife and family for their support throughout the course of this project.

Leonid Lerner
Flinders University
Adelaide, South Australia

About the Author

Leonid Lerner graduated with a BSc (Hons.) in 1984 and an MSc in 1985 from the Victoria University of Wellington, and obtained his PhD in physics from Cambridge University, UK, in 1989. From 1990 to 1993 he was a postdoctoral fellow at the Optical Science Centre and Laser Physics Centre of the Australian National University (ANU), and a lecturer in physics at ANU from 1994 to 1996. Dr. Lerner has worked in a number of different fields in the natural sciences and has published in a range of scientific disciplines.

Currently he is a research scientist at the Defence Science and Technology Organisation, Adelaide, and Honorary Fellow at the School of Chemical and Physical Sciences, Flinders University, Adelaide.

1 Safety in the Laboratory

1.1 INTRODUCTION

With today's emphasis on safety, even chemicals such as sodium carbonate and copper sulfate have been classified as hazardous, which can lead to complacency when dealing with the chemicals synthesized in this book. The significant danger these substances pose to the experimenter is best highlighted by the fact that their properties have necessitated restrictions on their mode of transportation; many are sea-freight items only.

Material safety data sheets (MSDS) are available on many Web sites—for instance, at http://msds.chem.ox.ac.uk—and the reader is advised to consult these. However, MSDS are generally generated from compilations of a chemical's properties, and the author has found, when synthesizing a hazardous chemical for the first time, that more specific information is often desirable. A good example is provided by triphosgene $(COCl_2)_3$ (Chapter 17). While this chemical is labeled "water-reactive liberating toxic gas" in the MSDS, one finds that it is actually essentially inert with respect to cold water, and thin coats deposited on glassware during preparation remain unreacted in water for many days. However, rinsing such glassware in acetone immediately releases phosgene, because acetone acts as an intermediate solvent between triphosgene and water. Gloves exposed to triphosgene vapor during preparation and touching the skin much later can produce red burn patches, whereas this does not readily occur with many other substances capable of causing skin burns, such as thionyl chloride, due to their much higher vapor pressure.

The dangers associated with the syntheses can be greatly reduced by a few simple items of safety:

- An effective fume hood
- Nitrile gloves
- Lab coat
- Eye-enclosing goggles

Dangerous situations nevertheless still arise due to unexpected problems, something which inevitably happens with sufficient time spent working in the laboratory. Examples of such situations are a pressure buildup due to blockage of glassware, equipment failure, or human error in manipulating chemicals and/or equipment. Protective equipment, such as a fume hood window, should never be bypassed during an experiment to rectify a problem. If an experiment gone wrong cannot be remedied without exposing the experimenter to danger, all power and gas generation should be turned off and the experiment commenced again when everything has cooled and the problem has been remedied.

All experiments described in this book have been carried out by the author several times, and when conducted as described, should not put the experimenter at risk. However, as a contingency measure, the hazardous substances prepared in the text are individually listed below, together with observations of their specific characteristics and hazards, situations under which these hazards may arise in the present syntheses, and methods that can be used to deal with them.

In addition, Table 1.1 presents a concise summary of the hazardous substances described in the book, using data obtained from MSDS literature [1].

1.2 CHARACTERISTICS OF HAZARDOUS SUBSTANCES PREPARED IN THIS BOOK

1.2.1 SODIUM

This is a reactive, very malleable metal with the consistency of cold plasticine, making it easy to cut with the end of a dry spatula. In air, sodium is covered in seconds by a gray oxide coat, which after several minutes turns white due to the formation of sodium hydroxide on reaction with atmospheric moisture. Other than in the form of a fine suspension, sodium does not usually spontaneously ignite in air, and the protective coating of paraffin under which it is often stored can be safely removed by washing with dry diethyl ether, toluene, or a similar hydrocarbon, without danger of fire. After long-term storage in paraffin whose surface has been exposed to air, a yellow solution, probably containing the superoxide, develops; sodium stored under such a solution maintains its luster for many years.

When dropped into water, sodium reacts, rapidly melts, and floats on the surface, releasing hydrogen gas. The greatest risk from this reaction is with quantities greater than about half a gram, which explode in air after a small time delay due to the accumulation of a substantial amount of hydrogen above the sodium, sending out a spray of sodium hydroxide solution and fine sodium particles.

In the sodium hydroxide cell, sodium at 320°C exposed to the atmosphere does not burn due to a very thin oxide layer that rapidly forms, and floats on top of it. If, however, it is deposited at this temperature in a thin liquid film, it catches fire with a delay of about 2–3 sec due to thermal runaway associated with reaction heating at a large surface-area-to-volume ratio. Such a situation applies to residues left in the ladle when sodium is removed from the cell. This residue can be rapidly destroyed by dipping the ladle into water held in a container surrounded by fireproof material. In this case, there is no explosion because, with the small amount of sodium, a substantial amount of hydrogen cannot build up. Sodium at 580°C in the Downs cell ignites instantly on contact with air and burns with a very bright yellow flame.

The molten sodium hydroxide electrolyte instantly dissolves most organic material, including any exposed skin. It is also very corrosive to most metals by acting as a descaling agent, leading to rapid oxidation. Molten sodium hydroxide baths release a corrosive mist of fine droplets when being dehydrated or electrolyzed, which are hazardous by inhalation. The greatest danger to the operator occurs when a sodium hydroxide cell is operated too close to the melting point (this should not happen if

TABLE 1.1

Major Hazards of Substances Prepared or Used in This Book

Substance	Toxicity	Flammability	Reactivity
Aluminum bromide	ORL-RAT LD$_{50}$ 1600 mg kg^{-1}		Water, acids
Bromine	IHL-MUS LC$_{50}$ 750 ppm/9 h		Metals, reducing agents
Calcium aluminum hydride			Water, acids, oxidizing agents
Carbon disulfide	ORL-RAT LD$_{50}$ 3188 mg kg^{-1}	Extremely flammable Autoignition 90°C	Oxidizing agents, azides, common metals
Carbon tetrachloride	ORL-RAT LD$_{50}$ 2350 mg kg^{-1}/ carcinogenic		Strong acids and bases, alkali metals
Cesium	ORL-RAT LD$_{50}$ 2600 mg kg^{-1} (CsCl)	Highly flammable	Water, acids, oxidizing agents
Chlorine	IHL-RAT LC$_{50}$ 293 ppm/1 h		Metals, reducing agents
Chlorosulfonic acid			Water, most metals, strong bases, sulfides, cyanides
Diethylaluminum bromide		Spontaneously flammable	Water
Hydrazine hydrate	ORL-RAT LD$_{50}$ 129 mg kg^{-1}/ carcinogenic	Explosion limits: 3.4–100% Autoignition 280°C	Oxidizing agents, dehydrating agents, metal oxides
Hydrazine sulfate	ORL-RAT LD$_{50}$ 601 mg kg^{-1}/ carcinogenic		Nitrites, oxidizing agents
Lithium	ORL-RAT LD$_{50}$ 526 mg kg^{-1} (LiCl)	Highly flammable	Water, acids, oxidizing agents
Lithium aluminum hydride			Water, acids, oxidizing agents
Lithium/Sodium hydride			Water, acids, oxidizing agents
n-Butyllithium		Spontaneously flammable	Water
Phosphorus oxychloride	IHL-RAT LC$_{50}$ 32 ppm/4 h		Water, many metals, amines, DMSO, strong bases
Phosphorus pentachloride	ORL-RAT LD$_{50}$ 660 mg kg^{-1}		Water, amines, sulfur, alkali metals
Phosphorus trichloride	IHL-RAT LC$_{50}$ 104 ppm/4 h		Water, organic materials, reducing agents

TABLE 1.1 (CONTINUED)
Major Hazards of Substances Prepared or Used in This Book

Substance	Toxicity	Flammability	Reactivity
Potassium	ORL-RAT LD$_{50}$ 2430 mg kg^{-1} (KCl)	Highly flammable	Water, acids, oxidizing agents
Potassium hydride		Highly flammable	Water, acids, oxidizing agents
Potassium t-butoxide			Water, acids, oxidizing agents, halogenated hydrocarbons
Sodium		Highly flammable	Water, acids, oxidizing agents
Sodium/potassium azide	ORL-MUS LD$_{50}$ 27 mg kg^{-1}		Oxidizing agents, acids, heavy metals
Sulfur mono/dichloride	IHL-MUS LC$_{50}$ 150 ppm		Water, many metals, alcohols, amines
Sulfur trioxide/Oleum	Highly toxic/ carcinogenic		Organics, metal powder, water, cyanides
Thionyl chloride	IHL-RAT LD$_{50}$ 500 ppm/1 h		Water, most metals, alcohols, amines, strong bases
Triethylaluminum		Spontaneously flammable	Water
Triphosgene	ORL-RAT LD$_{50}$ >2000 mg kg^{-1}		Strong acids and bases, oxidizing agents, amines

Sources: University of Oxford, Chem. Dept. Database: http://msds.chem.ox.ac.uk/#MSDS; Sigma-Aldrich Corp.: http://www.sigmaaldrich.com/safety-center.html.

the instructions in the text are followed) with the formation of a solid crust above the electrolysis zone. This can lead to ejection of bath when trapped hotspots form, as well as forcing the sodium metal from the frozen cathode compartment to the anode, leading to explosions with the expulsion of hot bath and a molten sodium spray. The cell, therefore, should be operated in an empty fume hood with a bucket of sand nearby (burning sodium is not extinguished by CO_2), and the operator should wear a full face mask, thermal gloves, and a lab coat when approaching the cell.

1.2.2   POTASSIUM

This metal is more reactive than sodium, has lower surface tension, and resembles soft butter in consistency. It is also less dense, with molten potassium rising in paraffin. These properties make the metal considerably more difficult to coalesce and purify. Potassium oxidizes substantially even under paraffin, with a thin purple coat forming in a matter of minutes and eventually turning gray-black and thickening over the course of several months. Depending on the hydrocarbon composition and purity, some metallation/reduction can also occur, leading to discoloration of the hydrocarbon.

In air, potassium tarnishes immediately as it is being cut. A thick, white crust several millimeters in depth develops in a matter of minutes and, when quantities greater than about 0.5 g are left in air, oxidation is accompanied by formation of a yellow superoxide, melting of the metal, and significant heating which can lead to ignition. This tendency of potassium to spontaneously ignite makes work with large quantities of the metal hazardous under a flammable hydrocarbon such as toluene, particularly in the presence of dispersed potassium. Fires must be extinguished by a large volume of sand, not CO_2. Potassium floats in some hydrocarbons (sp gr 0.88 at room temperature), a situation that should be avoided as it increases the risk of fire.

Even small globules of potassium dropped into water can lead to an explosion, because potassium reacts with water much faster than sodium. It, therefore, should never be disposed of in this fashion. Potassium waste is best treated with a large excess of t-butanol, with which it reacts slowly, globules the size of a pea being converted to the butoxide in a matter of 10–30 min. The reaction with primary alcohols is too vigorous and can lead to ignition.

Potassium at 410°C in the potassium hydroxide cell forms a shiny reflective layer that covers the entire surface of the electrolyte. The layer fumes profusely, but no flame develops due to shielding of oxygen by the oxide fumes. A potassium hydroxide crust is softer than that of sodium hydroxide so that the possibility of bath ejection due to encrustation is reduced. The bath is, however, much more corrosive to metals than molten sodium hydroxide, with copper dissolved at a rate of ~0.5 mm/h compared to <10 µm/h in sodium hydroxide.

1.2.3 LITHIUM

This metal is considerably less reactive than either sodium or potassium and is significantly harder. Small pieces can still be cut with the tip of a spatula, but at substantial pressure. Lithium is very ductile, and when poured from the cell can form long, needle-like strands. In air, lithium tarnishes at a slightly lower rate than sodium, forming a black mixed oxide/nitride coat, which unlike that of sodium and potassium is protective and superficial, halting further oxidation in the absence of moisture. The protective storage of lithium is complicated by its being lighter than almost all hydrocarbons (ligroin being an exception). It, however, can be stored under paraffin in a full bottle, or covered with a tight-fitting steel mesh, which prevents it from rising to the surface.

Lithium at 420°C in the LiCl-KCl cell does not tarnish, with large globules protruding up to 1 cm above the bath surface, and remaining shiny due to a very thin layer of electrolyte adhering to the surface. However, similar to sodium, small lithium residues can ignite in the ladle since the protective chloride coating is no longer present.

Lithium chloride, as any soluble lithium salt, is toxic by inhalation or ingestion, producing both physical and cognitive impairment of the central nervous system by interfering with sodium ion absorption in the body. The lithium ion has a LD_{50} ~100 mg/kg in humans.

1.2.4 Cesium

This metal is liquid above 28°C and, moreover, shows substantial supercooling. The liquid can easily be drawn up by a pipette, which should be completely clean, dry, and preliminarily filled with inert gas, as any contact with oxygen immediately deposits a dark purple/black higher oxide (up to Cs_9O is known) that sticks to the glass. Cesium does not attack glass up to 700°C (Chapter 5). The cesium produced here has a golden tinge, due to minute traces of oxygen present on the glass walls that are very hard to remove. Cesium can be stored as a liquid layer or solid block under paraffin for several months; however, this is accompanied by the gradual formation of a dark purple/black precipitated oxide above the cesium surface. Such storage is thus best conducted in tubes with a large volume-to-surface area ratio.

Even nondisperse cesium is pyrophoric in air, however; in confined spaces ignition is not instantaneous. Moreover, the higher-oxide formed can offer sufficient protection to avoid ignition altogether. Thus, if it is desired to store the cesium under paraffin rather than in evacuated tubes, it is possible to dehermetize the apparatus and dip the cesium below paraffin without its igniting, as described in Chapter 5.

Dropping a small amount of cesium into water leads to an explosive reaction.

1.2.5 Lithium and Sodium Hydrides

Both these substances react with water vigorously, releasing hydrogen gas that may explode. They are relatively inert in dry air and can safely be stored in a desiccator. Lithium hydride, pulverized in the manner described in the synthesis of Chapter 6, forms a small quantity of airborne dust, which is not ordinarily invisible, and is capable of strongly irritating the throat if inhaled. The operation should therefore be conducted in a fume hood.

1.2.6 Aluminum Bromide

This covalent compound is more corrosive and reactive than the more commonly encountered aluminum chloride. It fumes profusely even in relatively dry air due to the formation of hydrogen bromide and aluminum oxide/hydroxide. The latter, being particularly dispersed, is easily absorbed by inhalation into the body, with the concomitant health risks associated with absorbed aluminum.

Aluminum bromide reacts violently with water, ejecting a spray of steam, hydrogen bromide, and alumina fumes. Thermal shock from the reaction can also crack glass vessels. Excess aluminum bromide is best destroyed by reaction with ethanol, which is substantially slower and generates less heat. After the mixture has cooled, the glassware can be washed out with water.

Aluminum bromide instantly carbonizes many organic compounds including THF and diglyme, precluding their use as solvents. It also attacks silicone grease slowly, which combined with the associated formation of alumina can lead to glass stoppers' seizing. It decomposes cold diethyl ether only very slowly, and this compound is hence used to dissolve it.

Skin contact with aluminum bromide leads to instant and severe burns; therefore, nitrile gloves should be worn (thin latex provides inadequate protection).

1.2.7 Lithium Aluminum Hydride

This compound is pyrophoric and represents an explosion hazard due to hydrogen released by reaction with water and many solvents from which water can be obtained by dehydration. The synthesis covered in Chapter 9 reduces this danger by forming it in situ in an ether solution. However, the solution is unstable, and if it is not quenched immediately, the solved lithium aluminum hydride decomposes over several days—faster at raised temperatures or with ingress of humidity—releasing hydrogen, which can lead to pressure buildup or an explosion.

Lithium aluminum hydride reacts violently with water and hence should never be disposed of in this way. This reaction is so energetic that a dangerous situation arises when an ether solution is transferred into a new glass vessel containing traces of moisture on the walls, leading to vigorous hydrogen evolution and boiling of the ether, with the attendant danger of ignition or explosion. In the present experiment, a small amount of solid is left in the reaction vessel at the end of the quench, which contains some unreacted hydride. This can be decomposed by the dropwise addition of ethanol (ethyl acetate is preferable for large amounts) until gas evolution has ceased.

1.2.8 Diethylaluminum Bromide and Triethylaluminum

Both of these compounds are more hazardous than aluminum bromide because, in addition to reacting explosively with water, they are volatile and inflame spontaneously in air. Moreover, much heat is evolved if the reagents are transferred to a flask from which oxygen has not been entirely displaced. Diethylaluminum bromide is somewhat less inflammable, in that it is possible by quick action to transfer it between flasks using a syringe or even an ordinary pipette. Triethylaluminum, however, always inflames as soon as the tip of a pipette containing it is exposed to the air, but the flame is extinguished once the pipette is lowered into a nitrogen-filled container. Syringe transfer of these compounds through septa is hazardous due to the possibility of spilled drops catching fire during the transfer. The proper way to transfer these reagents is via Schlenk lines, using positive nitrogen pressure (negative pressure can result in the ingress of oxygen). Should a fire arise, it can be extinguished with CO_2. Excess reagent can be decomposed by slowly adding t-butanol.

1.2.9 Hydrazine Hydrate

This compound is a nerve poison and a probable carcinogen. In the anhydrous form it is inflammable, forming explosive mixtures with air. The hydrated compound is produced in Chapter 11, where it is volatilized by escaping carbon dioxide. Most of the hydrazine is absorbed by an acidic scrubber; however, because it is only weakly basic, some remains airborne in the CO_2 flux. For this reason, an efficient fume hood or venting system should be used. Hydrazine has a light amine (fish-like) odor rather than that of ammonia, as sometimes stated. Detection of this odor should not be relied upon because, at this stage, exposure is already above admissible levels.

1.2.10 AZIDES

Soluble metal azides are potent poisons of comparable toxicity to the cyanides. Like the cyanides, they react with acids to produce a poisonous gas, hydrogen azide, which is also explosively unstable and forms explosive mixtures with air. Discussed in Chapter 12, potassium and sodium azides are formed as insoluble precipitates in a basic alcoholic solution, which reduces the danger of accidental absorption; nevertheless, gloves and a fume hood should be used and, as with all hazardous preparations, washing up should be carried out in the fume hood.

1.2.11 N-BUTYLLITHIUM 2M IN HEXANE

This intermediate is used in the synthesis of Schlosser's reagent in Chapter 13. While it is classified as spontaneously inflammable in air, it is considerably less so than the ethylaluminum compounds mentioned in Chapter 10. Thus, it can be easily transferred between flasks capped with septa using a syringe, with little danger of the tip of the syringe or even small spilled drops catching on fire. Small amounts (a few milliliters) can be decomposed with water.

1.2.12 POTASSIUM HYDRIDE

This is a white, insoluble ionic compound, which is produced as a suspension in hexane, together with the fine lithium hydride described in Chapter 13. It is pyrophoric in air as well as presenting an explosion hazard due to the hydrogen released. It, therefore, should be used in situ and fully quenched at the completion of the experiment. Water should not be used in quenching.

1.2.13 CARBON DISULFIDE

This compound is a nerve poison and has one of the lowest autoignition temperatures for a stable compound (90°C), so that it may ignite from sufficiently concentrated vapor contacting a hotplate. Although the odor of the very pure compound does not present sufficient warning of a dangerous concentration, the carbon disulfide synthesized in Chapter 14 has sufficient odor due to the presence of organic mercaptan and sulfide traces, to provide adequate warning. Although the synthesis is conducted at well above the autoignition temperature (260°C–300°C), no particular shielding of hot surfaces was found necessary since ice–water cooling, as well as air flux in the fume hood, reduces residual carbon disulfide vapor concentration to acceptable levels. Ignition also does not normally take place should dehermetization of the apparatus occur for a short period, due presumably to the influence of the high concentration of sulfur vapor in the reactor on the autoignition point. This synthesis must be conducted in a fume hood with a CO_2 extinguisher nearby.

1.2.14   Hydrogen Sulfide

This gas, produced as a by-product in Chapter 14, is as poisonous as hydrogen cyanide. Although its odor detection threshold is well below that at which poisoning occurs, the olfactory nerves are rapidly numbed by its action, and hence corrective measures must be taken should its odor be detected, as subsequent lack of odor cannot be used as an indication of safety. Due to its acidity, it is almost entirely absorbed in the NaOH scrubber, with the fume hood disposing of any residual traces. Hydrogen sulfide should not be detectable outside the fume hood during the experiment.

1.2.15   Carbon Tetrachloride and Chloroform

These chlorinated hydrocarbons are both heavy liquids with low boiling points that exert a poisonous action on the liver and kidneys, and both are probable carcinogens. They possess a heavy sweet odor characteristic of strongly chlorinated hydrocarbons, but the odor is not sufficiently strong (550 ppm odor threshold for CCl_4) to provide adequate warning of dangerous concentrations. These liquids do not support combustion; however, if combustion is maintained by some other body and an adequate supply of oxygen is available, phosgene is generated. This is particularly serious in the case of carbon tetrachloride (about 300 mg phosgene can be generated per gram CCl_4 at 300°C). Reaction with oleum or hot sulfuric acid also yields phosgene. In Chapter 16, sulfuric acid is used in the scrubber with carbon tetrachloride present in the reactor; hence, there exists the possibility of carbon tetrachloride being sucked into the scrubber under fault conditions. The two liquids do not react in the cold, but an intermediate safety bottle can be used as described in the experiment.

1.2.16   Triphosgene

This compound has already been mentioned in the introduction to this chapter as an example of the usefulness of specific safety information. Its main use is as a safer phosgene substitute; it is formally a trimer of phosgene, and readily decomposes to phosgene under chemical or thermal action. The decomposition is particularly violent in contact with strong acids and bases, and should be avoided. Triphosgene also decomposes somewhat on standing at room temperature, and the smell of phosgene can readily be recognized on opening triphosgene containers (although no significant vapor pressure is built up). Substantial amounts of phosgene are also released when triphosgene is dissolved in solvents containing traces of water such as acetone, although the chemical does not noticeably react with cold water directly due to poor miscibility. Therefore, washing up must be carried out in the fume hood. Above its melting point (80°C), triphosgene generates substantial vapor pressure (it sublimes at 130°C at oil pump vacuum), which can deposit traces of triphosgene on cooler parts of the apparatus. If gloves contact these, and at some later stage touch bare skin, red burn patches can form after a short delay. Provided only small quantities of triphosgene are involved, the burns are painless and disappear in a few hours.

At medium concentrations, phosgene does not produce immediate substantial physiological action, which, however, can manifest itself after a latent period of 4–24 h in the form of a life-threatening pulmonary edema. Phosgene is readily recognizable at low concentrations by a sweetish smell resembling green grass or artificial air freshener. The odor threshold lies above the medium-term exposure safety limit and cannot be used as an adequate indicator of danger for long exposures at low concentrations. To adequately protect against these, commercial phosgene detection strips, or strips made as described in Chapter 17, should be employed during the experiment.

1.2.17 Phosphorus Trichloride, Oxychloride, and Pentachloride

These compounds present a danger in the form of both poisoning by inhalation as well as by their corrosive action on organic matter and violent reaction with water. At room temperature the danger of poisoning decreases in the order above due to decreasing vapor pressure, with phosphorus pentachloride being a solid at room temperature. Nevertheless, all compounds are produced at temperatures above their boiling points in the experiments of Chapters 18 and 19, and hence dehermetization can lead to the release of saturated vapor, which poses an immediate danger to health. In particular, phosphorus oxychloride, which has a structural and functional similarity to phosgene, being lipid soluble and hydrolyzed slowly enough to penetrate deeply into the lungs, is associated with the same risks of delayed pulmonary edema. The odor of these compounds, which is reminiscent of hydrochloric acid, does not provide adequate warning as it takes a few seconds to register, after which sufficient irritation of the upper airways has occurred; it takes a while to wear off.

Both phosphorus trichloride and phosphorus pentachloride react instantly and vigorously with water-forming phosphorous/phosphoric acid and hydrogen chloride gas. Phosphorus oxychloride has a latent period where it forms an immiscible lower layer below water, followed by sudden initiation and vigorous reaction. Thus, all reactions with these compounds must take place in a fume hood.

1.2.18 Sulfur Trioxide and Oleum

These chemicals are perhaps the most corrosive of any synthesized in this book. Sulfur trioxide vapor, even at the low vapor pressures that exist above the solid, rapidly carbonizes most organic material, including polyethylene tubing and silicone grease. Invisible traces of organic material left on glassware develop into a carbon trace. The traces disappear with further contact as sulfur trioxide oxidizes carbon to carbon dioxide. The attack on silicone tubing is so severe that it cannot be used in contact with sulfur trioxide vapor. The attack on polyethylene tubing is superficial at low vapor pressures, and it can be used instead. Sulfur trioxide does not attack Teflon at up to 200°C, and hence can be used with Teflon-coated stirrers and stopcocks. In perfectly dry apparatus, such as exist in Chapter 20, sulfur trioxide vapor is invisible, and the liquid is transparent. However, it fumes profusely even in a relatively dry atmosphere (it has been used as a smoke agent) due to the formation of a sulfuric acid mist. This vapor has no odor but is extremely corrosive, and is categorized as a probable carcinogen. Sulfur trioxide has a narrow temperature range for the liquid phase at atmospheric pressure, 16°C–44°C; moreover, there is substantial hysteresis, so the liquid tends to supercool, while the melting point is raised substantially by minute traces of water, which significantly modifies the solid structure. For these reasons, there is a danger of tube blockage in experiments involving volatilization of sulfur trioxide, and narrow bore tubing should not be used.

Sulfur trioxide reacts extremely violently with water, with instant vaporization of the surface layer on contact, and formation of a fine sulfuric acid mist, which is difficult to condense. For this reason, absorption is always carried out with concentrated sulfuric acid.

Oleum, being a solution of free sulfur trioxide in sulfuric acid also evolves fumes of sulfur trioxide when its concentration is greater than a few percent. The corrosive nature of the liquid, particularly at strengths exceeding ~10%, is comparable to that of anhydrous sulfur trioxide, and should be treated accordingly. It goes without saying that all manipulations of these chemicals should be conducted inside a fume hood that is capable of containing all mist.

1.2.19 Thionyl Chloride

Formally, this compound is a mixed anhydride of sulfurous and hydrochloric acids, and it behaves accordingly, forming both acids in a strongly exothermic reaction with water. The reaction is not instantaneous, and drops of thionyl chloride in cold water initially sink to the bottom without evident reaction; however, a highly energetic reaction commences suddenly, and this delay can be a source of danger because it is accompanied by boiling of the thionyl chloride, and vigorous expulsion of SO_2. Thionyl chloride is lipid soluble and is readily absorbed through the skin, causing severe burns. It bears some functional similarity to phosgene and, hence, represents a substantial danger of poisoning by inhalation with possible delayed pulmonary edema. It does, however, irritate the mucous membranes much more strongly than phosgene due to substantial decomposition to HCl and SO_2, and has a strong chocking odor reminiscent of both gases. Thionyl chloride decomposes substantially on distillation (about 10%) and to a lesser degree even at room temperature, thus significant pressure can build up in closed bottles. Thionyl chloride is a milder dehydrating agent than sulfur trioxide; for example, it does not attack polyethylene or silicone grease. All manipulations with this compound should be carried out in a fume hood, and full face protection should be worn when the window of the hood is raised.

1.2.20 Sulfur Chlorides

These are heavy, viscous, water-reactive compounds with a pervasive smell resembling sewer gas. Contact with organic matter, particularly unsaturated compounds (olefins), should be avoided due to the possible generation of chlorinated thio-ethers, which pose a far greater danger. These compounds are relatively unstable even at room temperature, with sulfur dichloride in particular existing in equilibrium with a substantial chlorine partial pressure. Sulfur chlorides react with water generating vapors of hydrogen chloride and sulfur dioxide, accompanied by the deposition of colloidal sulfur. The reaction is, however, fairly moderate with the main danger posed by the corrosive gases generated.

1.2.21   CHLOROSULFONIC ACID AND PYROSULFURYL CHLORIDE

Chlorosulfonic acid is the more dangerous substance due to its extremely vigorous reaction with water, which can result in the expulsion of boiling liquid and evolution of copious fumes of hydrogen chloride and sulfuric acid spray. Pyrosulfuryl chloride is the anhydride of chlorosulfonic acid. It is almost completely unreactive with cold water, but reacts moderately with hot water, yielding the same products as the acid.

Both substances are absorbed through the skin and cause severe burns. Danger by inhalation at room temperature is unlikely due to the high boiling points; however, the apparatus noted in Chapter 21 is saturated with the vapors of both substances, and this can present a danger if dehermetization should occur. For these reasons, the same precautions as for sulfur trioxide apply.

REFERENCE

1. University of Oxford, Chemistry Department Database. http://msds.chem.ox.ac.uk; Sigma-Aldrich Corp., http://www.sigmaaldrich.com/safety-center.html.

2 Sodium

SUMMARY

- Sodium hydroxide is electrolyzed at 315°C to produce sodium at near-theoretical yield.
- Cell theory is derived and compared with experiment.
- Sodium purity >99% by IC analysis.
- Sodium production ~11 g/h with a 50 A 4–9 V power supply.
- Cell requires attention from the operator every 15 min.

APPLICATIONS [1]

- Preparation of alkanes via the Wurtz reaction.
- Birch reduction of aromatics [2].
- Carbonation by formation of organometallic intermediates [3].
- Bouveault-Blanc reduction of esters.
- Preparation of organometallic compounds [4].
- Drying of solvents with sodium benzophenone.

2.1 INTRODUCTION

Sodium was first isolated by Sir Humphry Davy in 1807 in a series of experiments investigating the action on matter of the then newly discovered phenomenon of electric current [5]. A small amount of wet caustic soda placed between two platinum electrodes and heated by an applied current rapidly passes from the aqueous to the molten phase, which is then electrolyzed to produce small globules of sodium metal. However, it wasn't until 1891 that William Castner patented a commercial process based on this reaction, having discovered that Davy's method only works continuously if the melt temperature is held about 10°C, and no more than 20°C, above the mp of the caustic [6]. His patent consisted of an apparatus for realizing this fairly difficult requirement, when 64–75% of the supplied electrical energy is dissipated as heat in the sodium hydroxide cell, and the electrolyte is a fairly poor thermal conductor compared to sodium metal. His solution included the use of a bottom-side cathode, so that the sodium rises out of the hot electrolysis zone due to the large density differential between molten sodium and sodium hydroxide (sp gr 0.817 and 1.70 at 320°C, respectively), and thick solid electrodes to provide the necessary thermal conduction as well as an iron mesh separating the cathode and anode compartments. The mesh is an essential and fastest-wearing component of the cell, its pore size is a compromise between large pores needed for good circulation to avoid unacceptable temperature gradients in the bath and small pores that provide effective physical separation of the cathode and anode products.

In the laboratory, one typically electrolyzes dry molten NaOH in a nickel or steel crucible with a current of a few tens of amperes. This rapidly produces a small globule of sodium, provided the melt temperature is less than about 400°C. The globule initially sticks to the negative electrode due to surface tension and electrical forces; it does not burn as it is rapidly covered by a thin film of semitransparent oxide. However, after the first few hundred milligrams have formed, production of sodium ceases, and the amount already formed slowly starts dissolving. If electrolysis is continued further, the electrolyte darkens due to dissolved sodium attacking the cell, and the cell passes current without any evident chemical changes; the electrodes are then said to be depolarized. Addition of water in a stoichiometric amount to the dissolved sodium can restore cell operation but at reduced current yield. Moreover, if the cell is operated in the zero-yield regime for an extended period, operation can only be restored by complete renewal of the electrolyte. The occurrence of these phenomena in commercial cells is described by Wallace [7]; however, there is no detailed theoretical explanation in the literature, due largely to the fact that, with a few exceptions, the Castner process has long ago been superseded by the Downs cell, which electrolyzes sodium chloride. A theoretical framework of cell operation is presented below for the first time.

2.2   DISCUSSION

2.2.1   Basics of Cell Operation

Elemental sodium, which is produced by the reduction of sodium ions at the cathode

$$Na^+ + e^- \rightarrow Na, \tag{2.1}$$

is deposited in three phases: a condensed liquid, a metal fog [8], and a solution in the melt.

The following processes can occur at the anode:

$$2OH^- + \rightarrow H_2O + \tfrac{1}{2}O_2(g) + 2e^- \tag{2.2a}$$

$$Na \rightarrow Na^+ + e^- \tag{2.2b}$$

$$Ni + 2OH^- \rightarrow NiO + H_2O + 2e^-. \tag{2.2c}$$

Depending on the state of the cell, the following reactions can occur in the bulk electrolyte:

$$Na + H_2O \rightarrow NaOH + \tfrac{1}{2}H_2(g) \tag{2.3}$$

$$2Na + O_2(g) \rightarrow Na_2O_2 \qquad (2.4)$$

$$Na_2O_2 + 2Na \rightarrow 2Na_2O \qquad (2.5)$$

$$Na_2O + H_2O \rightarrow 2NaOH. \qquad (2.6)$$

Only Processes 2.1 and 2.2a are desirable redox reactions, with an electromotive force (emf) at the mp of pure NaOH (318°C) estimated to be 2.25 ± 0.05 V [9–11]. It also has been found that water, formed in Process 2.2a, is sufficiently soluble in NaOH at 318°C [11]; and that it almost completely reacts with sodium in Process 2.3, and hence the current efficiency does not exceed 50%. Depending on cell conditions, Le Blanc and Brode [12] describe Process 2.3 as occurring in either the anode or cathode compartments. Its location, which as explained below is critical to cell operation, can be looked upon as the interface between the acidic (excess water) and basic (excess sodium) regions in the cell. This correspondence follows from an acid/base equilibrium set up by dissolved sodium reacting with NaOH to produce Na_2O in the form of the highly basic anion O^{2-} [13] (which obviously cannot exist in appreciable concentration in aqueous solution being a much stronger base than OH^-)

$$2OH^- \rightarrow H_2O + O^{2-}, \ pK_{NaOH, \ 320°C} = 9.9, \qquad (2.7)$$

so that solutions with $pH_2O < \frac{1}{2} pK$ are defined acidic and vice versa. As the rate of sodium dissolution increases, hydrogen evolution, and therefore the acid–base interface, shifts to the anode compartment, thus commencing yield loss Processes 2.4–2.6. Depending on the convection pattern in the cell, Process 2.4 can take the form of sodium penetrating the mesh and reaching the anode in bulk as globules, leading to the formation of sodium peroxide by an explosive reaction with oxygen. Alternatively, sodium peroxide can form by oxygen reacting with dissolved sodium diffusing from the cathode. In both cases, sodium depolarizes the anode eventually halting oxygen evolution [9–11]. As the sodium peroxide concentration at the anode increases, so does the amount reaching the cathode by diffusion. This acts to depolarize the cathode by Process 2.5, so that eventually production of all material in the cell ceases, but it continues to pass current.

It is often stated that Processes 2.4–2.6 are a consequence of excessive sodium dissolution at high temperature [9–11], but as will be seen below, dissolved sodium concentration becomes excessive and will eventually halt the cell at any operational temperature if sodium is not removed from the cell by a certain characteristic time.

2.2.2 CELL REGIMES

When current is first passed through a neutral (dry) NaOH melt, sodium forms at the cathode without hydrogen evolution, and the cell functions in Regime 1:

Regime 1

$$2NaOH \rightarrow 2Na + H_2O + \frac{1}{2}O_2(g) \qquad (2.8)$$

However, within about a minute, depending on cell geometry, diffusion of water from the anode establishes sufficient water concentration in the cathode compartment for the cell to operate in Regime 2:

Regime 2

$$2NaOH \rightarrow 2Na + H_2(g) + O_2(g) \qquad (2.9)$$

This occurs because at the NaOH melting point, the rate of water evaporation from the cell is much less than its rate of diffusion from anode to cathode. Support for the large time-scale of evaporation is provided by the fact that an initially wet (acidic) NaOH melt containing ~1% H_2O requires about an hour at ~50°C above the mp for the water concentration to drop sufficiently for sodium, rather than hydrogen to be reduced at the cathode.

There have been various attempts to increase the evaporation to diffusion ratio, for instance, by operating the anode at far above the mp, or by employing semipermeable separators such as alumina screens [7,9,10]. These methods have produced cells operating in a mixture of Regimes 1 and 2, but the greater complexity increases cell attrition without increasing energy efficiency. For a normal cell, transition to Regime 2 is inevitable. However, extended electrolysis in Regime 2 leads to Processes 2.4–2.6, and so a transition to Regimes 3 and 4, which is now explained.

Von Hevesy [14] demonstrates that sodium exhibits substantial solubility in molten NaOH (25% at 480°C); however, solubility decreases with temperature, so this in itself cannot explain the loss of yield. Indeed, the absolute solubility of sodium in NaOH near the mp is immaterial to cell operation because it is much larger than the concentration at which sodium will be oxidized at the anode in preference to the hydroxide ion, and the cathode and anode reactions become as opposites of each other as follows:

Regime 4a

$$Na^+ + e^- \rightarrow Na \text{ (Cathode)} \qquad (2.10)$$

$$Na \rightarrow Na^+ + e^- \text{ (Anode)} \qquad (2.11)$$

In Regime 4a, current flows by means of a sodium ion flux from anode to cathode, with material sodium conservation being maintained by a flow of elemental sodium the other way, driven by a concentration gradient set up in the cell. In order for this regime to be achieved, sufficient elemental sodium must diffuse or convect to the anode to enable the cell current to be carried entirely by sodium ions. Any shortfall is made up by oxidation of the hydroxide ion, in which case the cell operates in a mixture of Regimes 2 and 4a.

From the diffusion Equation 3.1 in Chapter 3, the diffusion flux is proportional to the product of the diffusion coefficient and the concentration gradient, and it is the combination of these two quantities that determines when a cell starts to fail. The diffusion coefficient depends on temperature; in particular, sodium diffuses slowly near the mp of NaOH, and this is the origin of the stringent temperature requirement of the Castner patent. The concentration gradient in the steady-state is proportional to the total amount of sodium dissolved in the electrolyte. For a given electric current, there exists a threshold of dissolved sodium concentration at which the sodium ion flux sustaining the current can be entirely balanced by elemental sodium flux the other way. It is this concentration, rather than the saturated concentration of sodium in NaOH, which determines whether a cell operates in Regime 4. Convection is another transport mechanisms present in the cell, and similar arguments apply to it.

The author has observed that Regime 4 is preceded by a mode of cell operation with zero sodium yield but with gas evolution in the electrolyte bulk (but not at the anode). When sufficient sodium has dissolved so that the acid/base boundary is at the anode, the dissolved sodium unites with the free oxygen produced there, depolarizing the anode by Process 2.4. (Note that Na_2O cannot be produced at this stage as an overall oxidation reaction must occur at the anode to balance the hydrogen evolved by the cell). The cell now generates net sodium peroxide but no net sodium:

Regime 3

$$2NaOH \rightarrow Na_2O_2 + H_2(g) \qquad (2.12)$$

Provided the temperature is kept low, the cell can be maintained in this regime for an extended period; indeed, the commercial Castner process produces a substantial amount of Na_2O_2 by-product, which was marketed as a bleaching agent [7]. The same reference also mentions cell operation in alternate regimes of Na and Na_2O_2 production, but incorrectly associates the latter only with high electrolyte temperature, rather than with the amount of dissolved sodium.

As is the case with dissolved sodium in Regime 4a, the ultimate concentration of Na_2O_2 in Regime 4b is limited, since Na_2O_2 is produced by oxidation of sodium at the anode, whereas at the cathode Na_2O_2 inhibits sodium generation by the depolarizing action of Process 2.5 and 2.6. The net result is that an equilibrium concentration is reached in which sodium that is oxidized to Na_2O_2 at the anode is balanced by Na_2O_2 reduced at the cathode in the exact reverse reaction:

Regime 4b

$$Na_2O_2 + 2H_2O + 4e^- \rightarrow 2Na + 4OH^- \text{ (Cathode)} \qquad (2.13)$$

$$2Na + 4OH^- \rightarrow Na_2O_2 + 2H_2O + 4e^- \text{ (Anode)} \qquad (2.14)$$

Both Regimes 4a and 4b thus correspond to a state in which the cell continues to conduct current with no net chemical changes. Furthermore, the electrolyte in Regimes 4a and 4b is substantially basic, which is more corrosive to cell materials

Fe, Cr, and Ni by several orders of magnitude, as described by Estes [15]. Thus, with prolonged operation in this regime, the electrolyte becomes polluted with impurities from the corroding cell walls, which further increases the rate of sodium dissolution, so that replacement of the electrolyte is essential to resuming cell operation. In Regime 2, the cell corrodes more gradually, and the author has found that Process 2.2c accounts for about 0.5% of all electrochemical processes, with an anode loss of about 20 g Ni (or 0.3 mol) after the passage of 30 mol equivalents of electricity (neglecting corrosion in the absence of electric current). The dissolved metal oxides adhere to the sodium/melt interface, enhancing the rate of sodium dissolution.

Thus, it is seen that a key parameter on which the success of the Castner process depends is the rate of sodium dissolution in the melt near its mp. Indeed, it appears the extreme sensitivity of the yield to temperature, as well as very small concentrations of impurities, is due to the strong influence these exert on the dissolution rate. Thus, it has been reported by Wallace [7] that if the crust of solid electrolyte thrown by effervescence onto the walls of the sodium collector is returned to the melt, arcing due to the presence of sodium at the anode results, although the metal oxide concentration in the crust is barely measurable. Moreover, a concentration of calcium ions of less than 0.1% in the electrolyte can completely halt production of sodium, while small quantities of silicon deposited at the cathode can have the same effect [16,17].

2.2.3   CELL EQUATION

2.2.3.1   Preliminaries

It seems reasonable to suppose that the dissolution rate of sodium is proportional to the surface area of molten sodium A in contact with the melt:

$$\frac{dm}{dt} = -\beta A(t) \qquad (2.15)$$

where m is the mass of liquid sodium, and the proportionality constant β is a function of temperature, the composition of the bath including dissolved impurities and surface-active agents, and the flow regime in the bath, such as convection currents, etc. The surface area, in turn, is not a unique function of the amount of sodium present, but depends also on the extent to which the sodium is fragmented as well as the shape of the sodium layer as it is influenced by varying material and electric currents in the melt. The influence of the last two factors is not particularly significant for the following reasons. First, the surface tension of sodium is sufficiently large (about 180 mN/m as opposed to 104 mN/m for potassium at 320°C [18]), that it rapidly forms a single layer when agitated by convection currents in the bath. Second, one finds that for sodium globules in NaOH larger than a few millimeters, the globule height is almost independent of volume as characteristic of many sessile drops [19], so that the sodium surface area is proportional to its amount, and hence independent of the shape of the globule to first approximation.

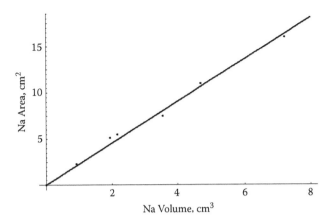

FIGURE 2.1 Optically measured surface area of sodium globules of varying shape and size floating on the surface of molten sodium hydroxide at 318°C, as a function of sodium volume. The nearly linear relationship demonstrates that the average globule height is almost constant.

To check this proposition, images of sodium globules of various sizes, produced in situ by electrolysis of an NaOH melt (containing 4% Na_2CO_3) so as to approximate as much as possible conditions of interest, were optically analyzed to determine the surface area. Figure 2.1 shows that to within the error inherent in such experiments for sodium globule sizes of interest, the sodium surface area is well fitted by a linear relationship to its volume V:

$$A = 2.26V \tag{2.16}$$

with the area in units of cm^2, and volume cm^3, respectively. Thus, the average globule height under static (no electrolysis) conditions is 4.4 mm.

This relationship would not be valid if the liquid sodium completely covered the collector surface. This is never allowed to happen in commercial cells [7,16], and even more so in laboratory-sized cells, as the collector surface area needs to be sufficiently large to allow the escape of evolved hydrogen without substantial turbulence or else the dissolution of sodium would greatly increase.

2.2.3.2 Relation between Drop Volume and Surface Area

Equation 2.16 is in good agreement with results obtained from the theory of surface tension. A drop of liquid A in hydrostatic equilibrium on the surface of a denser liquid B exhibits a balance of horizontal forces at the periphery of the drop [19]:

$$\gamma_B - \gamma_A \cos \theta_t = \gamma_{AB} \cos \theta_b, \tag{2.17}$$

where γ_A and γ_B are the surface tensions of the free surface of the respective liquids, γ_{AB} is the interfacial tension of the two liquids, and θ_t and θ_b are, respectively, the

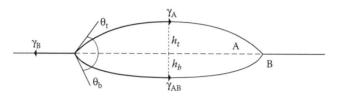

FIGURE 2.2   A globule of substance A floating on a denser substance B. The force balance determines the globule shape and height $h_t + h_b$. The height is nearly constant for globules whose width substantially exceeds their height.

angles that the tangents to the top and bottom drop surfaces make with the plane surface of liquid B, as shown in Figure 2.2. In many instances, when A exhibits considerable solubility in B, the interfacial tension γ_{AB} is small and can be neglected. Patrov and Yurkinskii [20] quote the density and surface tension of liquid sodium hydroxide at 400°C as 1.71 g/cm³ and 150 mN/m, respectively, which together with the surface tension of 180 mN/m for sodium quoted above, gives a contact angle of 34° with the horizontal.

A large drop is one whose gravity predominates surface tension, so that it is flattened with its top surface nearly horizontal. The horizontal surface forces on such a flattened section of width l are $(\gamma_A - \gamma_B)l$, and by reference to Equation 2.17 they clearly do not balance. The difference, ΔF, is made up by the horizontal force due to hydrostatic pressure, which is obtained by integrating the pressure ρgx a distance x below the surface of drop A, over the inclined face of the drop:

$$\Delta F = \frac{\rho_A l h_t^2 + (\rho_B - \rho_A) l h_b^2}{2},$$ (2.18)

where h_t and h_b are the heights of drop A above and below the surface of liquid B, respectively. Applying Archimedes law, and solving for the thickness of the drop $h = h_t + h_b$, we find [21]

$$h = \sqrt{\frac{2(\gamma_A - \gamma_B)\rho_B}{g(\rho_B - \rho_A)\rho_A}},$$ (2.19)

so that $h = 3.8$ mm for sodium in the melt, which is in reasonable agreement with observations.

2.2.3.3   Deriving the Cell Equation

Combining Equation 2.15 with the proportionality in Equation 2.16, the mass m and thus also the surface area A of sodium dissolving at a constant electrolyte temperature and composition should follow to first approximation an exponential decay with the same rate constant α,

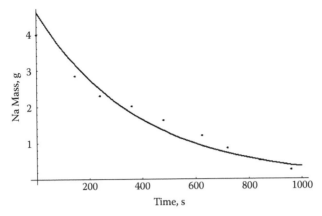

FIGURE 2.3 Variation in the mass of a sodium globule floating on the surface of molten NaOH with time from the moment of termination of electrolysis. The exponential fit provides support for Equation 2.20.

$$\frac{dm}{dt} = -\alpha m,$$ (2.20)

with α to be determined experimentally. Figure 2.3 shows the observed variation of sodium surface area with time from the moment of termination of electrolysis, with the electrolyte at an average temperature of 318°C, and demonstrates that the rate of sodium dissolution is in reasonably good agreement with Equation 2.20. Most of the small deviation between theory and experiment can be explained by ~ 4°C variation in temperature during the observations, its being difficult to maintain a constant temperature in the low-thermal conductivity melt during the transition from heating by electrolysis to external heating. In addition, one expects agreement to worsen when time becomes large because the shape of small globules approaches spherical, so that the considerations that led to Equation 2.19 break down, and A becomes proportional to $m^{2/3}$ rather than to m. From a fit of the data at these conditions, one obtains a dissolution rate of $\alpha = 0.0026\ \text{s}^{-1}$, applicable to Equation 2.20.

The rate of sodium generation at the cathode is by Faraday's law:

$$\frac{dm}{dt} = kI,$$ (2.21)

where I is the electrolysis current, and $k = M/F$; where M is the molar mass of sodium; and F is Faraday's constant. Equations 2.20 and 2.21 combine to give

$$\frac{dm}{dt} = kI - \alpha m.$$ (2.22)

This equation is only valid in Regime 1. Most useful sodium production occurs in Regime 2, when water has reached a steady state distribution in the electrolyte. Taking into account that in this regime at least half of the generated sodium dissolves by reaction with water diffusing from the cathode, we arrive at the *sodium cell equation*:

$$\frac{dm}{dt} = kI - \max(\frac{kI}{2}, \alpha m). \tag{2.23}$$

This equation has a finite $m(t = \infty)$ to which the mass tends at infinite time. Hence, if electrolysis continues indefinitely without removal of sodium, the percentage yield will approach zero, in agreement with the experiment.

Neglecting the processes of Regimes 3, 4a, and 4b, the mass of sodium dissolved in the bath n, plus the amount of bulk sodium m present at any given time, is the amount generated by Faraday's law minus the amount decomposed by water:

$$\frac{kIt}{2} = m + n. \tag{2.24}$$

We immediately see that a key difference between Equations 2.22 and 2.23 is that in the former case the amount of sodium dissolved in the electrolyte n continuously increases until eventually all current is carried by sodium ions and the cell becomes dead. In the latter case, no sodium concentration is developed in the electrolyte until time $t = 1 / \alpha$ corresponding to yield m_c:

$$m_c = \frac{kI\alpha}{2}, \tag{2.25}$$

which is proportional to the current, and equals half $m(t = \infty)$, the maximum yield that can be obtained from the cell. After this point, $n \neq 0$, m starts saturating, while n saturates even faster due to dissolved sodium oxidized at the anode by Processes 2.2b, 2.4, and 2.5.

We thus see that the diffusion of water, rather than being detrimental, is essential to the indefinite operation of the NaOH cell, *provided sodium is removed from the cathode at regular intervals, commensurate with $1/\alpha$*. If this is not done, the amount of bulk sodium starts to saturate, the sodium starts dissolving, and the cell commences functioning in Regimes 3, 4a, and 4b. This explains the low percentage yields reported by von Hevesy [14], where NaOH was electrolyzed for periods of the order of an hour at low currents.

Equation 2.23 does not describe the eventual disappearance of bulk sodium from the cell because it does not take into account that in Regime 4 part of the elemental sodium flux balancing the sodium ion current is carried by sodium already dissolved in the electrolyte rather than sodium generated at the cathode.

This reduces the effective current I in 2.21, and thus by Equation 2.25, also the maximum amount of sodium m_c the cell can support. Eventually n becomes sufficiently large that the ionic sodium current is entirely balanced by the material flux from sodium already dissolved in the electrolyte. This reduces the effective current to zero, and with it m_c.

The variation of the mass of liquid sodium m with electrolysis time, Figure 2.4, is in good agreement with Equation 2.23. The mass m was estimated by applying Equation 2.16 to time-lapse images captured during electrolysis. Because the current I varies with time during the experiment, kt in Equation 2.23 is replaced by the charge Q. The graphs show that initially m varies linearly with a charge passed through the cell, and starts saturating at large times. The agreement with theory is good despite the dissolution rate α used in the theory being measured under static conditions in the absence of material and electric currents.

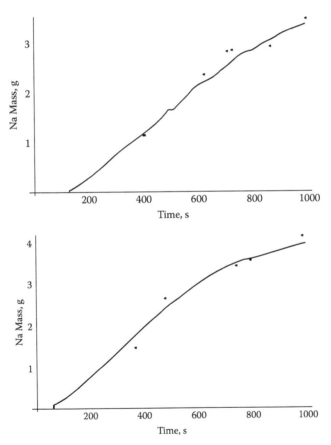

FIGURE 2.4 Variation of mass of liquid sodium m generated in the electrolysis of molten NaOH as a function of charge passed through the cell. The solid line represents data calculated from the cell Equation 2.23, and shows good agreement with experimental results, in particular demonstrating that the amount of sodium starts to saturate at large times.

2.2.4 Thermal Balance

An understanding of thermal balance is essential due to the stringent temperature requirements of the sodium hydroxide cell. This cell is quite wasteful of energy. As is typical for electrolysis, Processes 2.1 and 2.2a are accompanied by energy loss due to overvoltage and cell resistance. However, there are also substantial losses, Processes 2.3–2.6, taking place irreversibly, releasing heat into the electrolyte, with Process 2.3 occurring at the same rate as the main redox reaction. Moreover, the rates of Processes 2.4–2.6 greatly increase with temperature, making the cell subject to thermal runaway without an external negative feedback mechanism.

The reversible emf associated with Regime 1 is $\varepsilon_1 = 2.25 \pm 0.05$ V [9]; however, the cell actually operates in Regime 2, which differs from Regime 1 by the electrolysis of water, the emf that has been estimated at $\varepsilon_2 = 1.3$ V under cell conditions. However, in Regime 2, half of the NaOH decomposed reversibly at the electrodes is reformed irreversibly in the cell bulk by Process 2.3, depositing this energy as heat into the electrolyte. The total thermal energy deposited in the cell, neglecting the contribution of Processes 2.4–2.6, is then

$$\left(V - \frac{\varepsilon_1 + \varepsilon_2}{2} \right) I \approx (V - 1.8\gamma)I, \tag{2.26}$$

where γ is the current efficiency of the cell in Regime 1, and V is the applied voltage. As an example, in a typical industrial Castner cell, these quantities are 85–90% and 4.5–5 V, respectively [10], so that 64–71% of the electrolysis energy is deposited as heat in the cell, and must be removed to prevent the temperature rising.

The electrolyte itself has a rather low thermal conductance $\kappa \sim 9 \ 10^{-3}$ W K^{-1} cm^{-1} [22], compared to the copper cathode ~ 4 W K^{-1} cm^{-1}, and the nickel anode ~ 1 W K^{-1} cm^{-1}. Approximating a typical industrial Castner cell [10] by a cylinder, the thermal conductance of the cell is

$$C = \frac{2\pi k L}{\ln(d / d_1)}, \tag{2.27}$$

where $L = 30$ cm is the cell length, $d = 40$ cm is the cell diameter, and $d_1 = 10$ cm is the cathode diameter. We find $C \sim 1.4$ W K^{-1}, while thermal dissipation for a current of 1200 A in a typical Castner cell is 3.6 kW, so that the thermal conductance of the electrolyte in a Castner cell is inadequate to maintain thermal equilibrium, and convection is essential to prevent the formation of hot spots. This explains the poor yield obtained for electrolysis conducted too close to the mp (within 1°C–3°C), as circulation is hindered by high viscosity and solidification of part of the electrolyte [7].

Thermal energy is generated in the electrodes both directly due to cell overvoltage, and by conduction and convection from the electrolyte. In a typical Castner cell (copper cathode $d = 10$ cm, nickel anode $d = 18$ cm), these provide a combined thermal conductance of ~ 15 W K^{-1}, which is adequate to maintain thermal equilibrium with the cell operated ~ 300°C above room temperature. This number in practice is somewhat modified by the state of the electrode surface, and the conditions needed to avoid melt solidification on the electrode, which is a real possibility for an electrolyte close to its mp. The latter is determined by the thickness of the boundary layer near the electrode where conduction effects predominate convection, and is variable due to formation of an oxide coating as electrolysis proceeds. One can easily establish, using Equation 2.27, that no solidification results for a boundary layer less than 100 μm thick, and a bath 10°C above the mp.

2.2.5 TIME REQUIRED FOR MELTING

The low thermal conductivity of the electrolyte also manifests itself in the lengthy time needed to achieve complete electrolyte fusion, which for an oven set at 60°C above the mp and a crucible diameter of 12 cm, was found to take a little under 2 h. Neglecting surface effects, this situation can be modeled by the time required to fully melt a solid sphere of electrolyte initially at the mp, with its outer shell at radius r_0 held at temperature T above the mp. Neglecting convection, the energy balance at the melt front, with approximate account of electrolyte heating, is:

$$\frac{dr}{dt} = \frac{W}{4\pi r^2 \rho (\Delta H_f + TC_p)}, \tag{2.28}$$

where W is the power flux into the electrolyte, ΔH_f is the latent heat of fusion, ρ is the electrolyte density, and C_p is its specific heat capacity. The temperature distribution in the electrolyte satisfies Laplace's equation, $\nabla^2 T = 0$, which gives an expression for W in terms of T

$$\frac{W}{4\pi\kappa}\left(\frac{1}{r_0} - \frac{1}{r}\right) = T(r) - T. \tag{2.29}$$

Substituting Equation 2.29 into Equation 2.28, and solving for the time t_1 at which the melt front reaches the center of the sphere at $r = 0$, we find

$$t_1 = \frac{r_0^2(\Delta H_f + TC_p)\rho}{6T\kappa}. \tag{2.30}$$

Substituting $\rho = 2.1$ g cm^{-3}, $\Delta_f H = 160$ J g^{-1}, $C_p = 2.15$ J g^{-1} K^{-1}, we find $t_1 = 6700$ s, which is in good agreement with the observed time of 7000 s, considering we have neglected convection, the contribution to the thermal resistance from an oxide layer on the cell surface, and deviations from spherical symmetry.

2.2.6 CELL CORROSION

Cell corrosion cannot be regarded as a side issue in the sodium hydroxide cell because it directly influences the yield through the sensitivity of the sodium dissolution rate to the presence of small amounts of impurities in the electrolyte. Molten NaOH produces so-called hot corrosion, which is defined as accelerated corrosion occurring when a metal surface contacts a film of fused salt in an oxidizing environment [23]. By acting as a descaling agent, NaOH rapidly removes oxide layers from metals whose oxide has a negative solubility gradient in the oxide/salt/ O_2 interface (corresponding to oxide precipitation), in this case corrosion proceeds unabated [24]. In addition, corrosion occurs at a much faster rate during electrolysis because this generates reactive species such as water, and the oxidizing agents O_2 and Na_2O_2. The *ASM Specialty Handbook* [25] reports the corrosion rates of various steel compositions in molten sodium hydroxide, showing that stainless steels and carbon steels are about equally attacked in static tests. The author has found that this result also holds during electrolysis, with perforated S/S and carbon steel cylinders placed between anode and cathode shedding nearly equal masses of oxides in the electrolyte after a fixed time interval. Operation in Regimes 4a and 4b, which are most corrosive to the cell, gave a corrosion rate of about 6 μm/h at 30 A for both types of steel, which is much enhanced over the static tests, as expected. This is in accordance with Estes [15], who demonstrates an increase in the dissolution rate of both Cr_2O_3 and Fe_2O_3 of several orders of magnitude in molten NaOH at basicity (Na_2O content) of 10^{-7} mol/L. For this reason, nickel is preferential for anode and cell construction because among common metals it is the least affected by molten NaOH [15,23]. The interelectrode mesh is, however, most often made of steel, and since it is located in a high concentration of reactive species, it undergoes the greatest wear of all cell components [7,9].

2.3 EXPERIMENTAL

2.3.1 INTRODUCTION

From the foregoing discussion, it is apparent that with an average current of 40 A from a typical low-voltage laboratory power supply, 4–6 g of sodium will be generated before it has to be removed from the cell. Because this process is not easily mechanized, electrolysis is not a continuous process in the laboratory and needs to be interrupted every 10–15 min. It follows from this that the complexity of an underside electrode, used in the Castner cell to achieve continuous operation, is not justified in a laboratory setup. Moreover, a topside electrode allows the cell to operate in surface contact mode, with the cathode skimming the surface of the electrolyte, so the pool of sodium acts as the cathode. This has the benefit of reducing the surface area of sodium exposed to the electrolyte at the cathode to a minimum, while the enforced contact with the water-cooled copper electrode lowers the temperature of the sodium mass. This is very important in a small cell where hindered circulation due to a constrained space produces localized hot spots in the sodium collector typically 10°C or so above average electrolyte temperature. The copper electrode is 3/16 in. in diameter, and is cooled 2 cm above the melt surface by contact with a copper U-tube carrying water at room temperature. With a temperature difference of about 250°C across the electrode, this arrangement provides about 75 W of cooling, or about half of the total thermal power applied to the cell.

This arrangement works because the surface tension of sodium holds it in a single drop, which is attracted to the copper electrode by electrical forces acting to minimize electric field energy. The situation is analogous to electrostatics, where the electric field satisfies similar equations to those in an ohmic conductor, viz. $\nabla.\sigma\mathbf{E} = \mathbf{0}$ and $\nabla.\varepsilon\mathbf{E} = \mathbf{0}$. Electrical forces act on the mobile sodium pool in the same way they do on mobile electric charge, forcing it to spread to minimize electrical field energy, and preventing the formation of a narrow neck in contact with the electrode with increased resistivity and hence electric field intensity.

The longevity of the electrolyte is determined by the maximum allowable concentration of dissolved oxides. Because the rate at which these form is proportional to the surface area of the cell, longevity is proportional to the volume to surface area ratio. In a laboratory setup, this ratio is large, and to compensate for this the cell is made of nickel. This applies both to the cell walls and the cylindrical separator that makes a shallow dip into the melt, and separates anode and cathode.

It might appear at first sight that this arrangement fails, as the sodium pool can short the separator to the cathode so that sodium starts forming on the outer side of the separator. In practice this does not happen while the cell operates in Regime 2. Hydrogen gas generated by the reaction of anodal water with the dissolving sodium evolves predominantly on the separator surface. This evolution is especially intensive at the point of closest approach of the sodium pool to the separator, and acts to prevent contact. It also serves the useful purpose that when contact with the separator is finally made, the acid/base boundary has moved outside the cathode compartment, indicating that sodium must be removed from the cell.

Figure 2.5 shows the nickel separator enclosing the cathode compartment and the copper electrode attached to the sodium pool, which is stretched toward the anode by the **E**-field, and prevented from contact with the separator by hydrogen evolution.

2.3.2 SETUP AND OPERATION

A 500–1000-mL nickel crucible with 10–12 cm top diameter is filled with 800–1000 g 99% commercial NaOH, placed inside a crucible oven, and heated to a temperature of 360°C–380°C, as indicated by a thermocouple placed inside the melt. It is best to protect the thermocouple by a close-fitting copper sheath, which can be made of a sealed section of 3/16-in. copper tubing. This is best located in the anode compartment directly behind the nickel separator where the temperature registered is not prone to spurious variation due to intermittent contact with molten sodium. The NaOH takes about 2 h to melt fully, whereupon 40 g of dehydrated Na_2CO_3 is added (this serves to reduce the sodium diffusion rate, α), and the melt is held for an additional ½–1 h at that temperature to almost complete dehydration. Thereupon, the oven is turned off and a nickel separator, about 7 cm in diameter lowered to extend about 1–2 cm below the melt surface. When the electrolyte has cooled to 318°C, an overhead cathode, at a voltage of 12–15 V with respect to the nickel crucible anode, is lowered so that its tip just touches the electrolyte, and current begins to flow. The high voltage produces

FIGURE 2.5 *A color version of this figure follows page 112.* The nickel separator dipping below the molten NaOH encloses the cathode and forms the acid/base interface in the cell, and the region where hydrogen is generated. In Regime 2, this prevents sodium from shorting the separator to the cathode. The cathode consists of liquid sodium attracted by electrical forces to a water-cooled copper electrode raised just above the surface of the melt.

instantaneous heating, preventing immediate formation of an insulating crust on the electrode.

As soon as the current exceeds 20 A, the voltage is wound back to an average operating level of 6.3 V at 318°C. There upon the cell voltage is computer-controlled to counteract thermal runaway. This is achieved by decreasing the cell voltage linearly as a function of temperature, so that at 316°C the cell voltage rises to 6.7 V, with a corresponding decrease for a temperature variation in the other direction. This control does not need to be precise timewise, and a digital output fed through a low-pass filter with a 10-sec. time constant has proved adequate. During this time the cell current should vary in the range 25–55 A.

After a period of 10–15 min, more sodium starts dissolving than is reacting with the diffusing water, and hydrogen evolution moves outside the cathode compartment. This is accompanied by a sudden surge in current above 60 A, which is used to trigger an alarm indicating sodium needs to be removed from the cell. The cathode is lifted out, and a strainer (mesh 30/32 is adequate) is used to lift the pool of sodium and empty it into a container of paraffin. This method is used in the original Castner patent, and works because sodium has a reasonably high surface tension compared to the electrolyte. Any electrolyte frozen on the surface of sodium globules can be removed with tweezers if the paraffin is gently warmed to about 120°C. Sodium can be coalesced under paraffin at a few degrees above its mp of 98°C; at higher temperatures, it is too fluid. Removal of residual NaOH can be avoided if the strainer is fitted with a 1–2-cm steel cylinder on which the mesh rests. In this case, any residual NaOH in the strainer freezes to the walls of the cylinder when the sodium is poured. Although a small amount of sodium is lost this way, the recovered sodium is clean and ready for use (Figure 2.6).

The sodium should be removed from the cell as quickly as possible (the whole operation should take no more than about 30 sec, so that the melt surface does not have time to freeze), and the cathode tips are cleaned off any adhering solidified electrolyte by pressing between the jaws of small pliers so that the solidified melt flakes off. The cathode can now be reinserted, touching it against the electrolyte to commence the flow of current. When a globule of sodium has formed after about 30 sec or so, the cathode is raised 1–2 mm, and the whole process repeated. It should not now be necessary to raise the voltage above the normal level on reinsertion because the cathode is still hot. To prevent excessive freezing of the electrolyte surface during sodium removal, which can lead to spitting and small explosions upon recommencement of electrolysis, the oven is set to turn on when the bath temperature dips below

FIGURE 2.6 *A color version of this figure follows page 112.* Sodium as it is collected from the cell and stored beneath liquid paraffin.

314°C, which occurs only for a few minutes following sodium removal. After 80–100 g of sodium has been collected, fresh sodium hydroxide needs to be added (sodium carbonate is not added as it is not electrolyzed).

Although it is often suggested that prior to use sodium be cleaned from the adhering oxide/hydroxide crust by remelting and rolling under xylene [1], the evaporation rate of xylene above the mp of sodium has been found to be too high, and xylene is not very effective in removing the oxide crust. On the other hand, the bp of xylene is still high enough to present some removal problems, and it is unsuitable for long-term sodium storage because sodium rapidly oxidizes under it. In this respect, paraffin oil is superior. Not only does sodium retain its luster for months when stored under paraffin, but by heating the paraffin to above 120°C a light crust cover disappears in a matter of minutes, due to saponification of the paraffin and dissolution. To remove thicker crusts, the sodium surface can be trawled with fine stainless steel mesh. The paraffin adhering to the sodium surface cannot be volatilized as it has a very high boiling point, and some of its constituents carbonize without vaporizing even under a vacuum of 10^{-3} torr. However, paraffin can be fairly effectively removed by wiping the sodium with an absorbent lint free cloth, which still leaves a very thin layer of paraffin adhering to the sodium surface, preventing rapid oxidation in air during the wiping. The sodium is then shaken for several minutes with a large quantity of dry ether, which removes the last thin paraffin coat.

REFERENCES

1. Furniss, B. S., Hannaford, A. J., Smith, P. W. G., and Tatchell, A. R., *Vogel's Textbook of Practical Organic Chemistry, 5th ed.* London: Addison Wesley Longman, 1989.
2. Birch, A. J., Reduction by dissolving metals. Part I. *J. Chem. Soc.* 117: 430–436, 1944.
3. Moody, C. J. ed., Synthesis: Carbon with two attached heteroatoms with at least one carbon-to-heteroatom multiple link, pp. 27–8. Vol. 5 of *Comprehensive Organic Functional Group Transformations*, edited by Katritzky, A. R., Moody, C. J., Meth-Cohn, O., and Rees, C. W. New York: Pergamon, 1995.
4. Mordini, A., Sodium and potassium. In *Main Group Metal Organometallics in Organic Synthesis*, edited by McKillop, A. New York: Pergamon 2002; Jenkins, J. W., Nalo, L. L., Guenther, P. R., and Post, H. W., Studies in silico-organic compounds. VII. The preparation and properties of certain substituted silanes. *J. Org. Chem.* 13(6): 862–866, 1948; see also Chapter 10.
5. Davy, H., On some new phenomena of chemical changes produced by electricity, *Phil. Trans. Roy. Soc.* 98: 1–44, 1808.
6. Castner, H. Y., Process of Manufacturing Sodium and Potassium. U.S. Patent No. 452030, May 12, 1891.
7. Wallace, T., The Castner sodium process, *Chem. Ind.* 876–882, 1953.
8. Lorenz, R. and Clark, W., Über die Darstellung von Kalium aus Geschmolzenem Ätzkali, *Zeit. Elektrochem.* 9: 269–71, 1903.
9. Allmand, A. J. and Ellingham, H. J. T., *Principles of Applied Electrochemistry, 2nd ed.*, pp. 498–502. London: Edward Arnold & Co., 1924.
10. Thomson, G. W. and Garelis, E., *Sodium: Its Manufacture, Properties, and Uses*, pp. 20–23. New York: Reinhold Pub. Corp., 1956.

11. Mantell, C. L., *Industrial Electrochemistry*, *2nd ed.*, pp. 268–71. New York: McGraw-Hill, 1940.
12. Le Blanc, M. and Brode J., Die Elektrolyse von Geschmolzenem Ätznatron und Ätzkali, *Zeit. Elektrochem.* 8: 717–29, 1902.
13. Skeldon, P., Environment-assisted cracking of 2¼Cr-1Mo steel in fused sodium hydroxide at 623K, 1 atm—I. Electrochemistry in relation to stress corrosion cracking, *Corrosion Science* 26(7): 485–506, 1986.
14. Von Hevesy, G., Über die Schmelzelektrolytische Abscheidung der Alkalimetalle aus Ätzalkalien und die Löslichkeit Dieser Metalle in der Schmelze, *Zeit. Elektrochem.* 15: 529–36, 1909.
15. Estes, M. J., Corrosion of Composite Tube Air-Ports in Kraft Recovery Boiler: Cr_2O_3, Fe_2O_3, NiO Solubility in Molten Hydroxide. Ph.D. diss., Institute of Paper Science and Technology, Atlanta, Georgia, 1997.
16. Fleck, A., The life and work of Hamilton Young Castner. *Chem. Ind.* 515–21, 1947.
17. Hulin, P. L., Preparation of Sodium. U.S. Patent No. 1585715, May 25, 1926.
18. Jordan, D. O., and Lane, J. E., The surface tension of liquid sodium and liquid potassium. *Aust. J. Chem.* 18(11): 1711–18, 1965.
19. Couper, A., Surface tension and its measurement. In *Investigations of Surfaces and Interfaces—Part A*, edited by Rossiter, B. W. and Baetzold, R. C. New York: John Wiley & Sons, 1993.
20. Patrov, B. V. and Yurkinskii, V. P., Surface tension and density of a sodium hydroxide melt. *Russ. J. Appl. Chem.* 77(12): 2029–30, 2004.
21. Langmuir, I., Oil lenses on water and the nature of monomolecular expanded films. *J. Chem. Phys.* 1(11): 756–76, 1933.
22. Cornwell, K., The thermal conductivity of molten salts. *J. Phys. D: Appl. Phys.* 4: 441–45, 1971.
23. Rapp, R. A. and Zhang, Y. S., Hot corrosion of materials: Fundamental studies. *JOM* 46(12): 47–55, 1994.
24. Rapp, R. A. and Goto, K. S., The hot corrosion of metals by molten salts. In *Molten Salts 1*, edited by Braunstein, J. and Selman, J. R. The Electrochemical Society Proceedings Series, Pennington, NJ, 1981.
25. Davis, J. R., ed., *ASM Speciality Handbook: Nickel, Cobalt, and Their Alloys*, p. 185. ASM International, Handbook Committee, 2000.

3 Potassium

SUMMARY

- Molten KOH is electrolyzed at 410°C with a MgO separator, or reduced with Mg in paraffin, to produce potassium.
- Electrolyzed potassium yield is 20–25 g in ~50 min with current ~20 A.
- High current yield, 80–88%, due to surface contact mode.
- New theory shows the cell operates in a different mode to the sodium cell.
- Potassium purity >99% by IC analysis.

APPLICATIONS

- In situ preparation of potassium alkoxide, superoxide, amide, and butyl potassium [1].
- Forms stronger bases than sodium, for example, KNH_2 cyclizes propane [2].
- Scavenger of moisture and oxygen (more efficient than sodium) [3].
- Major component of NaK alloys, used as thermal fluids and to improve reactivity [3].

3.1 INTRODUCTION

This metal is a rarer reagent than sodium, and the method of producing it in the laboratory is considerably less well known. Its production by the electrolysis of molten KOH is an interesting demonstration of how a subtle alteration in physical parameters results in completely different cell behavior.

Although potassium was the first alkali metal produced by Sir Humphry Davy on electrolyzing molten KOH in the same fashion as he later repeated with sodium (he considered potassium production easier in that it could be achieved in thicker layers of KOH, and required a smaller current), an examination of his account reveals that his potassium was very impure, contaminated most likely with sodium. The melting point range of the substance he obtained was 10°C–21°C [4], corresponding to a NaK alloy of at least 21% sodium content, and is considerably below the 63.8°C mp of pure potassium [3]. This is not an insignificant fact because, as will be seen below, production of pure potassium by his method is impossible, and this led to confusion for those trying to repeat his work.

LeBlanc and Brode [5], attempting to repeat Davy's electrolysis, and following in particular Castner's patent, which claimed the cell could produce both sodium and potassium, found they could produce no potassium at all, save for some shiny streaks of metal that intermittently appeared and disappeared at the cathode. They attributed this failure to the great affinity of potassium for atmospheric oxygen, and concluded that Castner could not have produced potassium in his apparatus open to the atmosphere as he claimed, while pointing to the fact that Davy conducted his experiments with molten KOH inside glass tubes hermetically sealed with mercury. Davy's account, however, describes a set of experiments where he observed potassium globules forming at the bottom (cathode) of two platinum plates, subsequently rising to the anode through the melt, with the whole apparatus open to the atmosphere. The experiments in glass tubes were conducted subsequently to ascertain that atmospheric oxygen had no bearing on the observed results. LeBlanc and Brode also made the significant observation that KOH containing 10% water absorbs oxygen from the atmosphere when melting at 260°C, while the conductivity of substantially anhydrous KOH at 430°C increases by a factor of ~10 on being exposed to the atmosphere for a few hours, presumably due to superoxide formation.

Lorenz and Clark, on the other hand, attributed the failure of LeBlanc and Brode to the lack of cohesion of bulk potassium due to the formation of a metal fog in the electrolyte in a process they described as analogous to evaporation [6]. They considered potassium metal fog would form to a greater extent than sodium because it is closer to its boiling point at the mp of KOH (mp KOH 410°C, bp K 769°C) than sodium is (mp NaOH 318°C, bp Na 889°C). They succeeded in producing potassium in amounts of 10–15 g at current yields of up to 55% by using an inverted magnesium oxide crucible as separator between the anode and cathode compartments, which they considered acted to condense the potassium fog. Their MgO crucible had a small (3 mm) hole bored in its base to enable passage of the cathode, and when it had considerably filled with potassium it was almost completely withdrawn from the electrolyte while the latter was still viscous, and was allowed to cool with its mouth still immersed in the melt. It was subsequently broken under petroleum; the potassium was recovered, and the procedure repeated. Magnesium oxide is one of the few materials essentially inert to molten KOH and potassium, and a nonconductor, which is an essential property for a separator, as seen in Chapter 2.

Von Hevesy [7] in a set of experiments on the dissolution of K and Na in their hydroxides, demonstrated that above 480°C the dissolution of the metals in their hydroxides is real rather than colloidal in nature, in that to each temperature corresponds a saturation of solubility of the metal in the hydroxide, which obeys the ordinary solubility laws, including freezing point depression. His data showed that the solubilities are substantial (25 g for Na, 8.3 g for potassium, per 100 g hydroxide at 480°C), and that they decrease with temperature as expected, so that near its bp the solubility of potassium is only about 0.7 g. Von Hevesy also showed that potassium diffuses more slowly in its hydroxide than sodium (an analysis of his data gives a limit for the diffusion coefficient, $D < 3.8$ cm^2 h^{-1} in the range 320°C–700°C), but he reported some difficulty in determining the solubility and diffusion rate for potassium due to the fact that, rather than staying as a single mass, undissolved potassium was dispersed both as inclusions in the solidified hydroxide, and as layers on the container walls. The reason for this is described below.

At temperatures below 480°C, which are more relevant to electrolysis, he was unable to make measurements because of the low rate of diffusion. Von Hevesy was able to reproduce the potassium yields of Lorenz and Clark using an MgO separator (his KOH was substantially contaminated with NaOH judging by its mp); however, in light of the insignificant effect of metal fogs in his experiments, he agreed with Le Blanc and Brode that the main contribution of the crucible is the exclusion of atmospheric oxygen. Moreover, in a joint paper with Lorenz [8], he subsequently showed the contribution of metal fogs to potassium solubility is less than 0.1%.

We demonstrate below that Lorenz and Clark were essentially correct in that the main loss mechanism in KOH electrolysis is diffusion through the melt, rather than oxidation by the atmosphere; however, the formation of metal fogs is an inessential feature, and cell operation is explained here on the basis of the solubility von Hevesy observed, together with diffusion theory.

3.2 DISCUSSION

A molten KOH electrolyte was dehydrated as described in the experimental section and maintained at about 10°C above mp by temperature and current control, while being electrolyzed with a current of 10–20 A using the water-cooled copper electrode and nickel separator of Chapter 2. Shimmering streaks, suggestive of a very thin metallic layer, appeared and disappeared on the electrolyte surface in contact with the cathode, at the same time copious fumes of KO_2/K_2O_2 where evolved that coated the copper electrode. A strong current of argon gas was directed from a tube attached to the copper electrode onto the surface of the melt enclosed by the nickel separator, so that part of the electrolyte surface solidified; this halted oxide fume evolution, but had an insignificant effect on the longevity of the potassium streaks in the melt (Figure 3.1). Increasing either the electrical current or argon flow produced no visible effect. This strongly suggests that the main mechanism of potassium loss is not aerial in nature. If a small amount of potassium (corresponding to a charge of 0.1 M) is now produced by enclosing the cathode in an MgO crucible and the crucible then slowly raised, KO_2/K_2O_2 fumes begin to evolve from the electrolyte surface as the potassium-rich solution inside the crucible is expelled into the surrounding melt by hydrostatic buoyancy. When the crucible is sufficiently raised that liquid potassium is expelled, it immediately wets the entire surface of the electrolyte, producing a shiny metallic layer, which fumes strongly and disappears within a minute. A strong current of argon reduces the fuming, but does not significantly affect the rate of potassium absorption.

3.2.1 CELL EQUATION

3.2.1.1 Preliminaries

It is apparent from the above that the main difference between sodium and potassium in contact with their respective molten hydroxides is that the latter completely wets the surface of its hydroxide so that, provided the thickness of the potassium layer is

FIGURE 3.1 *A color version of this figure follows page 112.* Molten KOH at 410°C electrolyzed in the same arrangement as NaOH with currents in the range 10–50 A, produces fleeting metallic streaks in the vicinity of the cathode. Fuming is greatly decreased under a protective argon stream, but no potassium globules are formed, nor is oxygen effervescence observed at the anode.

substantially greater than molecular dimensions, its surface area is constant, independent of its amount, and is determined by the vessel in which it is contained.

Equation 2.17 shows that if the surface tension of the electrolyte is greater than that of the liquid metal, $\gamma_A < \gamma_B$, surface forces cannot balance and the metal spreads over the surface of the electrolyte completely wetting it. At 420°C the surface tension of potassium is about half that of sodium, $\gamma_A = 85$ mN/m [9], while that of potassium hydroxide is not available in the literature due to the difficulty of measuring this quantity in a high-temperature liquid corrosive to most metals. It can be estimated using a semiempirical formula derived by Reiss and Mayer [10], and shown to predict the surface tension of molten halide salts to within ~10% of the experimental value. It might be expected to apply to hydroxides as well, since the hydroxide ion behaves as a single atom anion in many respects (the O-H bond is short, and the hydrogen atomic radius is small compared to oxygen). Indeed, for sodium hydroxide with density $d = 2.07 - 4.78 \cdot 10^{-4} T$ [11], and minimum interatomic separation $a = 236$ pm (the same value as for NaCl because OH⁻ and Cl⁻ have nearly identical covalent radii), we find $\sigma = 155$ mN m⁻¹ at 320°C, in excellent agreement with 150 mN m⁻¹ quoted in Chapter 2. For KOH, with $d = 2.01 - 4.4 \cdot 10^{-4} T$ [11] and $a = 267$ pm, we find the surface tension $\sigma = 136$ mN m⁻¹ at 410°C, which substantially exceeds that of potassium. The observation that potassium completely wets the surface of its hydroxide is thus in accordance with theoretical results.

Hence, potassium does not form globules on the surface of molten KOH, which explains the difficulty von Hevesy encountered in his solubility experiments, while the globules Davy observed can be explained by the fact that he produced a NaK alloy, which has higher surface tension [12] than pure potassium. It is thus apparent that potassium cannot be separated from the melt in the same fashion as sodium.

Lorentz and Clark were able to obtain potassium in the condensed phase by restricting the surface area over which the potassium spreads from the cathode to the 3 cm diameter circular cross section of their MgO crucible. The crucible also hindered convection and increased the path length from anode to cathode by its 10 cm length, so that the considerable solubility of potassium was moderated by its low diffusion rate. Another circumstance limiting convection in the KOH cell is the fact that, unlike NaOH, no measurable hydrogen gas evolution is observed in the cathode compartment. The primary reason for this is the higher operating temperature of the KOH cell, which means that water evaporation dominates diffusion.

On the other hand, the author has found that NaOH electrolysis with an MgO separator leads to poor results, as indeed observed by von Hevesy. First, the solubility and diffusion rate of sodium are much larger than those of potassium, so that saturation of the surface layer on which the potassium cell relies is never attained with sodium, while the longer electrical path length increases thermal heating in the cathode compartment, which, combined with the very rapid increase in the diffusion rate of sodium with temperature [7], decreases yield. Second, it is observed here that the electrical path is very soon open-circuited by hydrogen pressure buildup, forcing the electrolyte out of the separator (a fact not mentioned in [7]), so the cathode compartment in a sodium cell cannot be well sealed.

These circumstances also explain why no potassium layer forms immediately when the melt is protected by argon gas. Potassium metal is produced at a finite rate on the melt surface in accordance with Faraday's law, but since it rapidly forms a thin layer of molecular dimensions it immediately dissolves and starts diffusing down toward the opening of the crucible driven by the concentration gradient. If the relationship of the rate of formation to the rate of diffusion of the potassium is such that at a certain time the concentration of dissolved potassium at the surface layer reaches saturation, at that moment potassium appears in the condensed phase and starts accumulating.

These ideas can be expressed quantitatively using the diffusion equation for the concentration C

$$\frac{dC}{dt} = D\frac{d^2C}{dx^2}, \tag{3.1}$$

where D is the diffusion coefficient, and the x axis points downward, with the origin located at the K–KOH interface. The domain is defined by the cylindrical crucible, and Equation 3.1 is one-dimensional due to the uniform concentration within the circular cross section, which is enforced by the spreading of the potassium in a uniform layer on the surface. The initial condition is

$$C(t = 0, x) = 0 \tag{3.2}$$

representing an initial zero potassium concentration inside the cylinder. The (mixed) boundary condition

$$C_x(t,x=0) = -\frac{I}{FAD\rho_M} \tag{3.3}$$

is obtained by setting the diffusion flux $J = -DC_x$ at the melt surface equal to the flux of potassium generated by the current passing through the cell in accordance with Faraday's law, $J = I/FA\rho_M$, where F is Faraday's constant, ρ_M is the molar density of potassium, and A is the cross-sectional area of the crucible.

Equation 3.1– Equation 3.3 have the following closed form solution over an infinite domain:

$$C(x,t) = \frac{I}{FAD\rho_M}\left(2\sqrt{\frac{Dt}{\pi}}\exp^{-x^2/4Dt} - x\operatorname{erfc}\left(\frac{x}{2\sqrt{Dt}}\right)\right). \tag{3.4}$$

This solution is not applicable at large diffusion coefficients $4Dt > L^2$, since it predicts a significant potassium concentration at the crucible mouth $x = L$ during electrolysis. In reality, the potassium concentration effectively drops to zero at the mouth, since it is consumed there by the bulk electrolyte, aided by the fact that the KOH melt is oxidative in nature, as demonstrated in Section 3.2.2. Hence, we have to enforce a zero potassium concentration by the following boundary condition, which makes the problem finite domain:

$$C(x = L, t) = 0. \tag{3.5}$$

The following solution to the finite domain boundary value problem, Equation 3.1– Equation 3.3 and Equation 3.5, does not have a closed form:

$$C(x,t) = \frac{IL}{FAD\rho_M}\left(1 - \frac{x}{L} - \frac{8}{\pi^2}\sum_{n=0}^{\infty}\frac{1}{(2n+1)^2}\cos\frac{(2n+1)\pi x}{2L}\exp^{-D(2n+1))^2\pi^2 t/4L^2}\right), \tag{3.6}$$

but for sufficiently large D, only the first few terms in the expansion are important. Setting the concentration on the surface of the electrolyte equal to saturation $C(0, t) = C_0$, we can solve for the time t_0 at which potassium first appears on the surface in the infinite domain solution, Equation 3.4, valid for small D,

$$t_0 = \frac{\pi D}{4}\left(\frac{C_0 FA\rho_M}{I}\right)^2. \tag{3.7}$$

For large D, the finite domain solution Equation 3.6 is applicable, and we approximate this by the first three terms to obtain the following expression for the *characteristic time of potassium condensation*:

$$t_0 = -\frac{4L^2}{\pi^2 D}\left[\ln\left(1 - \frac{C_0 FAD\rho_m}{IL}\right) + 0.21\right].$$ (3.8)

The three-term approximation breaks down when the current is greater than about three times the diffusion current.

The only unknown in these equations is the diffusion coefficient D, which we will first estimate from the static data of von Hevesy, obtained in the absence of ionic fluxes and convection. His diffusion experiments conducted over a period of 1 h over a distance of 3 cm, showed a potassium concentration of less than 0.5% w/w at all temperatures. The boundary condition for this diffusion problem corresponds to a saturated concentration of potassium C_0 maintained at the K–KOH interface at all times:

$$C(0,t) = C_0,$$ (3.9)

which gives the following solution to the diffusion Equation 3.1:

$$C(x,t) = C_0 \mathrm{erfc}\left(\frac{x}{2\sqrt{Dt}}\right).$$ (3.10)

The saturated concentration of potassium in KOH at 410°C is ~0.083 w/w [7], and using $C(t = 1\text{ h}, x = 3\text{ cm}) \sim 0.005$, we find $D \sim 1.27$ cm²/h.

If this value applies over typical electrolysis times $t \sim 1$ h, the characteristic diffusion length is $L = 2\sqrt{Dt} \sim 2$ cm which is less than the cell dimensions so that D is small, and Equation 3.4 can be used to determine the appearance time of potassium. Substituting $C_0 \sim 0.23$ for the v/v saturated concentration of potassium in KOH, as well as $A = 8$ cm² for the separator cross section and $I = 10$ A for the cell current, we obtain $t_0 \sim 32$ sec. This time corresponds to a diffusion length of only 1.0 mm, and since surface tension imposes a minimum distance of 2–3 mm to which the cathode must be submerged in the melt, there are significant errors associated with modeling potassium generation as a surface effect. If we model it as a bulk effect and assume that the potassium is generated uniformly along the length of the cathode L, that its concentration is uniform over the crucible cross section, and that it varies only in depth, the diffusion Equation 3.1 becomes

$$\frac{dC}{dt} = D\frac{dC^2}{dx^2} + \frac{I}{FAL\rho_M}H(L-x),$$ (3.11)

where $H(x)$ is the Heaviside step function, and the boundary condition is zero flux at the origin $C_x(0, t) = 0$, and zero initial concentration $C(x, 0) = 0$. The solution to Equation 3.11 on the surface of the melt is

$$C(0,t) = \frac{I}{FAL\rho_M} \int_0^t \text{erf}\left(\frac{L}{2\sqrt{Du}}\right) du. \tag{3.12}$$

This can be inverted as before to give the time at which the potassium concentration exceeds saturation. Assuming a submersion length of $L = 4$ mm gives $t \sim 128$ sec. considerably greater than in the surface approximation.

    The consequences of a substantial cathode immersion are much more substantial in the experiments of Lorentz and Clark and von Hevesy, who considered the utility of the separator only in the protection it offered from the atmosphere, and hence submerged the cathode the whole length of the crucible. This is clearly detrimental to the yield because, due to the resistivity of the bath, the potassium is generated predominantly in the section of the cathode close to the mouth of the separator where the diffusion flux out of the crucible is also greater. In addition, potassium rising to the surface creates convection currents that tend to saturate the entire volume of the crucible before potassium is deposited in the condensed phase. These effects, being convective in nature, are difficult to quantify; however, they account for the lower yields observed in these experiments (26% at 410°C) compared to our yields of 77–88% at 410°C, obtained with the surface contact mode.

3.2.1.2   Deriving the Cell Equation

Once the condensed potassium layer has formed, the problem becomes one of diffusion from the region of constant potassium concentration on the K–KOH interface into the electrolyte inside the crucible. Because we are now considering long times, the problem is finite domain, so that the initial condition and boundary conditions are

$$C(x = 0, t) = C_0, \tag{3.13}$$

$$C(x = L, t) = 0, \tag{3.14}$$

$$C(x,t = 0) = C_0\left(1 - \frac{x}{L}\right). \tag{3.15}$$

Equation 3.15 satisfies the diffusion Equation 3.1, and hence *is* the solution for all times. It corresponds to a linear decrease in concentration from C_0 at the K–KOH interface to zero at the mouth of the crucible. This general solution for the concentration field is now used to develop the cell equation.

    The location of the K–KOH interface in the cell normally shifts as potassium accumulates. If the gas inside the cathode compartment cannot escape or the crucible is submerged to its base, the distance of the interface $L(t)$ from the crucible mouth is

$$L(t) \sim L_0 - \frac{M(t)}{A\rho_M} \tag{3.16}$$

where $M(t)$ is the number of moles of condensed potassium present at time t. If the cathode compartment is not airtight, the location of the interface is determined by the buoyancy of potassium in the electrolyte, corresponding to a factor of 0.39 in front of the second term.

Now, the total amount of potassium dissolved is given by the integral of the flux from the K/KOH interface $J(t) = -DC_x(0,t)$ over the time t of cell operation. From Equation 3.15, which applies for $t > t_0$, we get a differential equation for the number of moles of dissolved potassium $M_d(t)$

$$\frac{dM_d(t)}{dt} = \frac{C_0 DA\rho_M}{L(t)} = \frac{M_0 I_d}{M_0 - M(t)} \qquad t > t_0, \tag{3.17}$$

where we have combined factors to define the diffusion flux I_d from the K–KOH interface at the moment potassium first appears

$$I_d \equiv \frac{C_0 DA\rho_M}{L_0}, \tag{3.18}$$

and

$$M_0 = AL_0\rho_M, \tag{3.19}$$

is the maximum number of moles of potassium a crucible submerged to depth L_0 holds.

Assuming that the cell current does not influence the diffusion coefficient as a first approximation, we equate the rate of change of the number of moles of condensed potassium plus the rate of potassium dissolution, Equation 3.17, to the current passing through the cell I (in moles) to obtain the *differential potassium cell equation*

$$\frac{dM}{dt} + \frac{M_0 I_d}{M_0 - M} = I \qquad t > t_0. \tag{3.20}$$

Solving Equation 3.20 using the initial condition $M(t_0) = 0$, gives us the *integral potassium cell equation*:

$$I(t - t_0) = M - M_0 \frac{I_d}{I} \ln\left(1 - \frac{MI}{M_0(I - I_d)}\right) \tag{3.21}$$

where t_0, defined by Equation 3.8, can be reexpressed in terms of I_d:

$$t_0 = -\frac{4C_0 M_0}{\pi^2 I_d}\left[\ln\left(1 - \frac{I_d}{I}\right) + 0.21\right]. \tag{3.22}$$

For small amounts of potassium and diffusion currents, $M \ll M_0$ and $I_d \ll I$, the logarithm in Equation 3.21 simplifies, and we obtain the simpler *reduced potassium cell equation*:

$$(I - I_d)(t - t_0) = M. \tag{3.23}$$

It is clear from Equation 3.21–Equation 3.23 that potassium condenses in the cell only if the current I exceeds the diffusion current I_d . The closer I is to I_d, the smaller the saturation level at which potassium must be cleared from the cell, $M = M_0(1 - I_d/I)$. This is always less than M_0, and hence potassium will never spill out of the separator.

3.2.1.3 Experimental Confirmation of the Cell Equation

A direct experimental confirmation of the cell equation is more difficult for potassium than for sodium, because visual detection of the amount of bulk potassium, M, is not possible. An approximate determination of M was made by weighing the potassium at the end of the experiment, which introduces a substantial error of about $+2 \pm 0.5$ g, since the potassium does not coagulate well in the KOH melt, as discussed above.

Table 3.1 lists the yield in grams of potassium as a function of time for various runs made with a constant total charge passed through the cell, $Q = 0.7$ mol, but with variable current, and hence electrolysis time t. The mean current could not be substantially increased beyond about 25–30 A without overheating the cell. The separator was submerged to its base $L = 6$ cm, giving $M_0 = 0.92$ mol. A best fit to the table data gives a diffusion flux of $I_d = 1.0 \ 10^{-4}$ mol s^{-1}, or $FI_d(0) = 9.6 \pm 2.5$ A. Using the relationship expressed by Equation 3.18 between the diffusion coefficient D and the diffusion current I_d, we find $D = 61$ cm^2 s^{-1}, which is more than an order of magnitude greater than that obtained under static conditions. The time at which bulk potassium appears in the cell is therefore quite large: $t_0 \sim 10$ min for a current of 20 A. *This explains why no potassium was observed in the experiment shown in Figure 3.1.*

Of particular note is the data in the last row of the table, which shows zero yield for a constant 10 A cell current and correspondingly long electrolysis time. This

TABLE 3.1
Variation of Potassium Yield with Electrolysis Duration at Charge $Q = 0.7$ mol

Time (sec.)	Mean Current (A)	Potassium Yield (g)	Current Yield (%)
2930	23	21	77
2938	23	24	88
3358	20	17	62
3422	19	19	70
3630	18.5	16	59
6418	10.5	0	0

cannot be explained by atmospheric effects and clearly demonstrates the presence of a potassium diffusion current. While yields of up to 55% have been reported by von Hevesy at currents of 10–12 A, these experiments are not directly comparable with those presented here. The cell of von Hevesy clearly contained a substantial NaOH impurity as it was reported to operate at 320°C–380°C, which is below the mp of pure KOH, and thus produced a NaK alloy. Moreover the diffusion current is proportional to the inverse of the distance between the mouth of the crucible and the K–KOH interface, which in the present case being 6 cm, is almost half that used by Lorenz and Clark.

3.2.2 CELL REGIME

From the high potassium yields at large currents in Table 3.1, and with diffusion theory explaining the decrease in yield at low current, it is apparent that the potassium cell operates for the most part in Regime 1, with the loss Processes 2.2b–2.6 representing a minor contribution. In particular, energy loss due to the decomposition of water, Process 2.3, which limits sodium cell efficiency to less than 50%, is very minor in the potassium cell. This is immediately apparent from the fact that the potassium cell for the most part of its operation does not generate any evident gas bubbles at the cathode; indeed, if this were not the case, a sealed cathode compartment could not be used, as demonstrated by the electrolysis of NaOH with a MgO separator.

Moreover, at anodic current densities normal for the sodium cell ($\sim$ 1 A/cm^2), the electrolysis produces almost no oxygen gas at the anode; consequently, the spray or small explosions that plague the sodium cell, are absent in the potassium cell. Of course, oxidation must occur to balance the evident reduction of potassium ions at the cathode, and the only candidate processes are the oxidation of nickel, Process 2.2c, and/or oxidation of hydroxide ions to the peroxide/superoxide (oxidation states -1, $-1/2$), with the latter process more energetically favorable:

$$4OH^- + K^+ \rightarrow KO_2 + 2H_2O + 3e^-. \tag{3.24}$$

Testing nickel anodes for changes in weight after lengthy periods of electrolysis has produced no evidence that Process 2.2c occurs at any more than a fraction of a percent. On the other hand, the magnitude of the cell potential for anodic production of superoxide in a KOH cell, calculated using the free energies of formation at 406°C (which neglects the solvation energy of the various species in molten KOH), is less than for the anodic production of molecular oxygen, suggesting the formation of superoxide is favored thermodynamically:

$$4KOH \rightarrow 3K + 2H_2O + KO_2 \quad \varepsilon_{T=406} = -2.9 \text{ V} \tag{3.25}$$

$$2KOH \rightarrow 2K + H_2O + \tfrac{1}{2}O_2 \quad \varepsilon_{T=406} = -3.2 \text{ V}. \tag{3.26}$$

The lack of stability of NaO$_2$ explains why Process 3.24 does not occur in the sodium cell.

While initially KOH electrolysis produces no evident O_2 evolution, after about 0.5–0.7 mol of charge has passed through an electrolyte containing about 23 mol KOH, a small amount of gas evolution at the anode becomes apparent at current densities exceeding about 1 A/cm². A substoichiometric evolution of O_2 at the anode at high current densities had also been reported by Le Blank and Brode [5], who also noticed the gradual absorption of oxygen gas from the atmosphere by a fresh KOH melt combined with an order of magnitude increase in its electrical conductivity (no such effects were observed in molten NaOH).

It is thus apparent that KOH dissolves O_2 significantly, and substantially more than NaOH, so that after exposure to the atmosphere the KOH melt forms an oxidizing medium. Nevertheless under standard conditions KO_2 is decomposed by H_2O, and indeed in the pure state it decomposes above 520°C, hence KO_2 formation in the electrolyte is in equilibrium with decomposition to O_2 and/or K_2O_2, as well as with the evaporation of H_2O from the surface.

The position of the equilibrium is of importance to cell operation, for if the KO_2 concentration were to build up substantially, corresponding to a long characteristic lifetime in the bath, cell operation would likely change substantially with time, and the yield of potassium decrease. Experiments to determine the concentration of KO_2/K_2O_2 species were performed by removing ~100 g (7%) samples of the bath after the passage of 0.7 mol of charge, cooling, quenching in water, and determining the amount of gas evolved. The latter was found to be less than 100 mL in all cases, the error being due to the large evolution of steam by the heat of hydration of KOH. This means that the amount of O_2 present in the cell after the passage of 0.7 mol of charge is less than 0.05 M, and shows that KO_2 equilibrium is very quickly established, after which time all formed oxygen and water is lost to the atmosphere at the surface. This effect does not produce noticeable effervescence at low currents since O_2 is generated effectively by KO_2 decomposition occurring in the entire electrolyte volume, rather than at the anode as in the NaOH case. Hence, after a small period of time the potassium cell operates in Regime 1, with Reaction 3.24 being the effective anodic process.

3.2.3 Cell Corrosion

The greater oxidizing power of the KOH melt compared to NaOH is easily observable [13]. Although Estes et al. [14] mention a factor-of-three increase in corrosion rates of stainless steels compared to NaOH, this is highly dependent on the amount of dissolved oxygen, and the author has found that corrosion rates for S/S immersed in the melt during electrolysis can be up to 85 μm/h, that is, 10 times greater than in NaOH. In contact with nickel, this increased to about 300 μm/h due to the galvanic effect. The corrosion rates for copper, which is much more sensitive to oxidation above 300°C, generally were of the order of 300 μm/h—except in the reducing environment of the cathode. This completely precludes the use of these materials in cell construction. In particular, thermocouples needed to measure electrolyte temperature must be encased in a nickel sheath. Surprisingly, nickel in KOH showed no observable increases in the corrosion rate over NaOH, the main observable difference being that in KOH the stoichiometric green NiO forms almost exclusively, whereas oxidation in NaOH produced mainly the nonstoichiometric black variety.

3.3 EXPERIMENTAL

The cell container and anode are similar to those in Chapter 2, with the cell being a nickel crucible of 500–1000 mL capacity and a top diameter of about 12 cm placed in a top-loading crucible oven, and the anode a 7-cm-diameter nickel ring of about 5 cm length suspended in the melt. The separator consists of a MgO crucible of 32 mm internal diameter, with walls 3 mm thick, and a length of 7.5 cm [15].

The cathode is of wider diameter than in the NaOH case, which counteracts somewhat the greater resistance of the KOH cell due to the MgO separator. The electrode is formed by an 8-mm-diameter copper rod, 3 cm long, brazed at the end to a 3.2-mm-diameter copper tube that passes through the base of the MgO crucible. A hole about 3.3 mm in diameter is drilled using a diamond tip in the base of the crucible. If a tungsten carbide tip is used instead of a diamond tip, small cracks in the base due to stresses have been seen to develop with thermal cycling; they do not, however, generally affect the performance of the separator. The hole diameter in the base is slightly larger than the copper tube it passes, being a compromise between allowing passage of gas to enable rapid equilibration of fluid inside the crucible, and restricting the access of atmospheric oxygen. The 3.2-mm copper tube is brazed, a distance about 2.2 cm above the crucible to a water-cooled copper U-tube of the same type used in the sodium cell. This configuration provides about 40 W cooling to the cathode, which proved adequate at currents up to 30 A.

Potassium hydroxide 99%, sourced as flakes or pellets, contains 10–12% water of hydration. This aqueous compound melts at about 260°C–280°C to a clear liquid of low viscosity, and it is found that almost complete dehydration, such as sufficient for the reduction of potassium ions at the cathode, can be achieved by soaking the melt in a crucible oven held at 380°C for 2 h, followed by 3–4 h at 450°C. With this procedure there is no bubbling, frothing, or effervescence reported by Lorenz and Clark [6]. The mp at the end of the dehydration rises to around 405°C. If one examines the heating/cooling time versus temperature curves of the anhydrous compound, an inflection (change in heat capacity) is observed at around 258°C, corresponding to a solid-state phase transition [16].

Upon completion of the dehydration, the cathode surrounded by the MgO separator is lowered into the melt, about 1 cm below the point at which the circuit is closed (Figure 3.2), which limits potassium diffusion from the electrode surface. If there are problems in closing the circuit due to gas trapped inside the separator, a snorkel can be formed from a small copper tube, inserted the whole length of the crucible during lowering and then removed.

With 99% KOH (excluding water of hydration), there were no initial explosions or crackling in the cell as reported by Lorenz and Clark [6], the electrolysis always proceeds quietly and smoothly for the entire duration. Variation in temperature is computer controlled, as in the sodium cell, by applying linear negative feedback with temperature to the voltage at the cell terminals: 412°C sets 9 V, 415°C sets 7.2 V, with the voltage varying linearly in between. The oven heater is controlled in an on/off fashion at 412°C. The above parameters limit the temperature variation during electrolysis to 3°C–4°C. The current is much more steady than in the sodium case, as the cross section of the liquid potassium does not vary with

FIGURE 3.2 *A color version of this figure follows page 112.* A 3-in.-long MgO crucible, which is not attacked by the caustic, is used to separate the anode and cathode compartments in the KOH cell. A water-cooled copper tube acts as the cathode and supports the crucible. Unlike the NaOH case, no hydrogen is generated in the cathode compartment; however, the copper–MgO juncture is not airtight to allow the fluid level inside the crucible to equilibrate.

time. The cell passes 20 A for most of the run, rising to 30 A at the end due to decreasing cell resistance as the K–KOH interface nears the separator mouth, and thus thermal dissipation decreases.

Breaking the MgO separator to extract the potassium [6] can be avoided by placing a 30–50 mL nickel or S/S crucible, whose lip is welded to a long rod, over the mouth of the MgO separator prior to raising it from the melt, so the potassium and residual KOH are held inside the separator by buoyancy. This arrangement is then transferred to a container flushed with argon and carrying a vessel containing paraffin oil, and the metal crucible quickly dipped below the paraffin surface. The separator need not touch the paraffin and can immediately be reinserted into the melt to start another run.

When the contents of the metal crucible have cooled somewhat, the lid of the container can be removed and any liquid potassium remaining in the crucible scraped into the paraffin. This procedure is not as simple as with sodium, since the potassium, which somewhat resembles soft butter in consistency, sticks to the solid KOH, as shown in Figure 3.3. Some loss of potassium as inclusions in the melt is inevitable.

Due to a much lower surface tension than sodium, the potassium gathered under paraffin is in the form of a grey/purple mass of very irregular shape, completely coated on the outside by oxides and covered with black inclusions of frozen hydroxide bath. No metal can be seen on the surface. The inclusions cannot be removed by

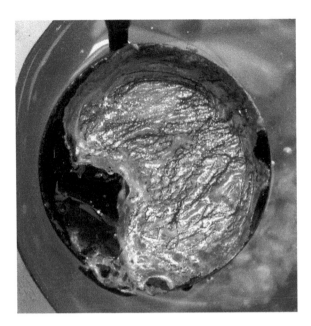

FIGURE 3.3 *A color version of this figure follows page 112.* When the contents of the metal ladle have cooled under paraffin, the potassium is scraped from the solidified KOH layer below into the paraffin-filled container. The potassium has a blue-violet tinge due to traces of oxygen, and resembles soft butter in consistency.

scraping the surface as with sodium, due to the low surface tension of the potassium that deforms on scraping into oxide-coated smaller globules, which are even harder to purify. Moreover the metal cannot be separated from the paraffin by distillation under vacuum (10^{-2} torr and lower) because it co-distills with some of the hydrocarbons in paraffin under these conditions. Use of low-boiling-point hydrocarbons such as toluene or petroleum ether to elute the paraffin is problematic since a small amount of potassium is present in the form of granules or powder, which are pyrophoric in air. Such emulsions occasionally spontaneously ignite and are difficult to extinguish. Adding a small amount of alcohol to the metal immersed in hydrocarbon for the purpose of removing the oxides [17], has been found to produce much occlusion due to saponification, but does not substantially aid in purifying or coagulating the metal.

However, several simple methods can be used to produce essentially pure metal. First, the density of potassium changes from about 0.89 g/mL at room temperature to about 0.83 g/mL at its melting point of 63.4°C, which is below the density of most paraffin oils at that temperature (0.85–0.89 g/mL), so that some potassium is able to free itself off the crust and float to the surface, where it can be collected with a strainer. The bath temperature must not be raised much above the mp of the potassium, since at higher temperatures the paraffin density drops again below that of potassium. Second, the potassium can be freed off the crust and oxides by use of a pipette (20–50 mL) with a 2–3 mm opening. Because at its melting point potassium floats in paraffin, the latter is replaced by lower density, low-flammability hydrocarbons such as Shellsol D70 (alkanes and cycloalkanes of average molecular weight 174 g/ mol, bp 200°C–240°C, density 0.79 g/L). A small amount of solvent is preliminarily drawn into the pipette to protect the potassium from air, the pipette is pressed 3–3.5 mm through the oxide coat into the molten potassium underneath, and gentle suction applied. Pure shiny metal is thereby drawn off and is then redeposited under clean solvent into buttons of the desired size, and left to cool. The oxide-covered "bag" thereby shrinks in size, and the process is continued until no more buttons can be collected without a substantial increase in suction, corresponding to the drawing of some of the oxides into the pipette. The remaining oxide-coated bag still contains some recoverable potassium, and this can be drawn by coalescing the bag with others of the same type, thereby decreasing the area-to-volume ratio. As with sodium, potassium globules can be run into one another by pressing them together only at a few degrees above the melting point, because at higher temperatures its viscosity is too low.

The potassium can be stored under paraffin or Shellsol D70 in tightly stoppered containers. Due to its greater reactivity than sodium, the potassium buttons lose their metallic luster in a few days and become coated with a purple/gray oxide.

Small amounts of potassium can be prepared in a surprisingly simple reaction [18], which is found here to be based on the following set of transformations in paraffins,

$$Mg + 2t\text{-BuOK} \rightarrow 2K + (t\text{-BuO})_2Mg,$$

$$KOH + (t\text{-BuO})_2Mg \rightarrow MgO + t\text{-BuOK} + t\text{-BuOH},$$

$$K + t\text{-BuOH} \rightarrow t\text{-BuOK} + \tfrac{1}{2}H_2,$$

Thus, 100 ml Shellsol D70, 12.6 g KOH, and 6.4 g magnesium powder (up to 80% can be turnings) are heated in a 250-ml conical flask, attached to a Liebig condenser, bubbler, and dropping funnel. At 150°C, the magnesium reacts with the molten hydrated KOH, generating hydrogen and MgO. Now 1.2 g t-BuOH in 5 ml D70 are slowly added and the temperature raised to 220°C, generating 5.6 g potassium (82% yield) in 4 hours. During the reaction the potassium is coalesced by the regeneration of alcohol and this can be completed by swirling the solids at 65°C in a stoppered flask with 50 ml dioxan to which 0.2 g t-BuOH have been added. The initiation stage is very sensitive to K-reactive species in the paraffin, and equivalent reactions with sodium are much slower due to the lower solubility of sodium alkoxides.

REFERENCES

1. Furniss, B. S., Hannaford, A. J., Smith, P. W. G., and Tatchell, A. R., *Vogel's Textbook of Practical Organic Chemistry, 5th ed.*, pp. 744–45. London: Addison Wesley Longman, 1989.

2. Tietze, L. F. and Eicher, T., *Reactions and Syntheses in the Organic Chemistry Laboratory*, Section 3.1.4. Sausalito, CA: University Science Books, 1989.

3. Chiu, K.-W., Potassium. In *Kirk-Othmer Encyclopedia of Chemical Technology, 5th ed.*, Vol. 20. New York: John Wiley & Sons, 2006.

4. Davy, H., On some new phenomena of chemical changes produced by electricity, *Phil. Trans. Roy. Soc.* 98: 1–44, 1808.

5. Le Blanc, M. and Brode J., Die Elektrolyse von Geschmolzenem Ätznatron und Ätzkali, *Zeit. Elektrochem.* 8: 818–222, 1902.

6. Lorenz, R. and Clark, W., Über die Darstellung von Kalium aus Geschmolzenem Ätzkali, *Zeit. Elektrochem.* 9: 269–71, 1903.

7. Von Hevesy, G., Über die Schmelzelektrolytische Abscheidung der Alkalimetalle aus Ätzalkalien und die Löslichkeit Dieser Metalle in der Schmelze, *Zeit. Elektrochem.* 15: 529–36, 1909.

8. Von Hevesy, G. and Lorenz, R., *Zeit. Phys. Chem.* A74: 443, 1910.

9. Jordan, D. O., and Lane, J. E., The surface tension of liquid sodium and liquid potassium. *Aust. J. Chem.* 18(11): 1711–18, 1965.

10. Reiss, H. and Mayer, S. W., Theory of the surface tension of molten salts. *J. Chem. Phys.* 34: 2001–3, 1960.

11. Janz, G. J., Thermodynamic and transport properties of molten salts: Correlation equations for critically evaluated density, surface tension, electrical conductivity, and viscosity data. *J. Phys. Chem. Ref. Data* 17(2): 109–159, 1988.

12. Lebedev, R. V., Measurement of interphase surface tension of sodium-potassium alloys. *Rus. J. Phys.* 15(12): 1825–27, 1972.

13. Bessey, C. M., Method for Descaling Metal Strip Utilizing Anhydrous Salt. U.S. Patent No. 5377398, Jan 3, 1995.

14. Estes, M., Marek, M., Singh, P. M., Rudie, A., and Colwell, J., Corrosion of composite tube air-ports in Kraft recovery boiler: Cr_2O_3, Fe_2O_3, NiO solubility in Molten hydroxide. Paper presented at the 9th International Symposium on Corrosion in Pulp and Paper Industry, May 26–29, 1998, Ottawa, Ontario.

15. A possible source is at http://www.graphitestore.com

16. Seward, R. P. and Martin, K. E., The melting point of potassium hydroxide. *J. Am. Chem. Soc.* 71(10): 3564–65, 1949.

17. Dönges, E., Alkali metals. In *Handbook of Preparative Inorganic Chemistry, 2nd ed.*, p. 965, edited by Brauer, G. New York–London: Academic Press, 1963.

18. Krennrich, O., Brendel, G., and Weiss, W. Process for Producing Alkali Metals in Elemental Form, U.S. Patent No. 4725311, February 16, 1988.

4 Lithium

SUMMARY

- Lithium is produced by electrolyzing LiCl/KCl in a near-eutectic mixture at 420°C.
- Current yield is ~100% in an apparatus open to the atmosphere with uncooled iron cathode.
- Single run produces 0.4–0.7 mol lithium at 6.5 V and 20 A.
- Lithium ~99% pure, with potassium < 0.1%, much lower than reported [1].

APPLICATIONS

- In situ preparation of organolithium compounds, for example, fresh *n,t*-butyllithium solutions [2].
- Birch (solvated electron) reduction in liquid ammonia [3].
- Synthesis of efficient hydrogen storage materials, for example, LiH, LiAlH$_4$, etc. [4].

4.1 INTRODUCTION

A favorable combination of physical and chemical properties makes lithium the least challenging of the alkali metals to produce electrolytically in the laboratory. Lithium is useful for the preparation of a series of organolithium compounds for although the Li–C bond displays a greater covalent character and less reactivity than that of other alkali metals, it is more stable. Organolithium compounds are stronger bases and nucleophiles and thus more reactive than Grignard reagents, with whom they share a number of common reactions.

The existence of lithium was first heralded in 1817 by the discovery of a base with a smaller equivalent mass than either sodium or potassium in the mineral petalite LiAlSi$_4$O$_{10}$; however, the properties of the metal could not be analyzed until Bunsen and Matthiessen isolated it in 1855 [5]. They electrolyzed the relatively low-melting LiCl (mp 606°C) in a cell with an iron cathode and graphite anode, and observed the formation of silver-white reguli the size of a small pea every 2–3 min, which were withdrawn together with the electrode with the aid of a small flat spoon and collected under oil. Hiller [6] enclosed the cathode in a clay pipe inverted in the LiCl melt, through which a stream of dry hydrogen was passed to protect against oxidation. Because molten lithium reduces silica very rapidly [7], the pipe had to be protected by a baked-on layer of powdered graphite dissolved in LiCl. The LiCl was preliminarily fused with a small amount of ammonium chloride to prevent the traces of water present in the deliquescent LiCl hydrolyzing the lithium ion to LiOH, with the subsequent attack of the hydroxide on the porcelain cell container. Common metals are unsuitable as cell materials because they are rapidly attacked by the hot chlorine gas convecting from the anode.

Electrolysis of pure LiCl produces variable results and generally poor current yields, and this was improved on by Guntz [8], who lowered the mp to 400°C with an electrolyte consisting of equal weights of lithium and potassium chloride (corresponding to a 63:37 molar ratio), which is 5% on the lithium-rich side of the 58:42 eutectic (with the eutectic melting at 352°C). By staying on the lithium-rich side of the eutectic, solidification is avoided soon after commencement of electrolysis due to decreasing lithium content. Guntz surrounded the 4 mm iron-wire cathode by a glass separator of 20 mm diameter, resulting in a high cell resistance, requiring a potential difference of 20 V to pass 10 A. After an hour of operation, the lithium inside the glass tube rises more than 1 cm above the melt level, and is scooped out of the melt with an iron spoon on termination of the electrolysis. It is presumed that long-term operation of such a cell is impractical due to lithium's attacking the glass.

Despite the lower decomposition potential of lithium chloride (3.68 V [9]) compared to potassium chloride, and the poor alloying ability of lithium with potassium, Roff and Johansen [1,10] found that the electrolytic lithium produced by the above method had a variable potassium content of up to 5%. They also observed that a pure LiCl electrolyte was a poor conductor requiring voltages of 33 V or more to pass currents of 20–30 A in a small cells, and this led to rapid deterioration of the carbon anode. They, therefore, trialed an electrolyte consisting of LiBr mixed with 10–15% LiCl with a mp of 520°C, which gave a current yield of around 80% at a current of 100 A and a potential difference of 10 V.

On repeating the above experiment, the author has found that already at 50 A the amount of chlorine gas evolved induces sufficient convection and effervescence to lose much electrolyte by spilling and spitting in the 8-cm-diameter 6-cm-deep cell used by Roff and Johansen, while without a separator the extreme convection disrupts the LiCl layer that normally protects the molten lithium, exposing it to the chlorine effervescence saturating the melt and resulting in frequent explosions and loss of yield. The graphite cell is also rapidly worn at current densities of ~ 3 A/cm^2, and this pollutes the bath. On the other hand, the small amount of potassium impurity mentioned by Roff and Johansen failed to materialize in the present experiments, with IC spectra of lithium samples quenched in water indicating a lithium purity >99%, with <0.1% potassium present (see Appendix, Figure A.3).

The electrolytic cell described below consistently produces current yields of ~100% within experimental error (±0.2 g). The electrolyte is lithium rich with respect to the eutectic, consisting of an equal mass LiCl–KCl mixture at 410°C–420°C, while the cathode uses an iron wire operated as a surface contact electrode, so that the formed lithium acts as the surrogate cathode. The cell container is a carbon crucible that also acts as the anode (resulting in low anode current densities ~0.15 A/cm^2). The percentage yield is independent of the duration of electrolysis and the amount of lithium produced (in the range 0.1–0.7 mol), provided the melt remains on the lithium-rich side of the eutectic. A surprising feature is the simplicity of the cell, which uses no cathode/anode separator or atmospheric protection. This is particularly fortunate as it is quite troublesome to find a suitable material to act as separator in a small LiCl cell. Common inert ceramics, such as alumina and magnesium oxide, are attacked by the chlorine [11]. Thus, the author has found that the MgO crucible used in the potassium cell rapidly deteriorates when used as a separator in the lithium cell. This is due to the dual action of anodal Cl_2 and cathodic lithium. The former leads to rapid pit corrosion in the basic MgO, while the latter reduces MgO to magnesium, quickly producing a dark layer at the point of contact.

4.2   DISCUSSION

Lithium exhibits substantial differences in its physicochemical properties from the other alkali metals, most of which facilitate the design of an electrochemical cell for its production. These characteristics include the following:

- Small ionic radius leads to lithium salts with large halogenic anions melting at substantially lower temperatures than those of other alkali metals (606°C for LiCl, 546°C for LiBr), contributing to a reduction in lithium diffusion, and lower reactivity at the electrolyte temperature.
- Lithium exhibits no noticeable solubility in its halide salts at their mp, so that anodic depolarization by the dissolved metal, which is so troublesome in the sodium cell, is effectively absent.
- Lithium has a higher melting point than those of the alkali metals, and an associated higher surface tension (365 mN m^{-1} at 400°C [6]), which aggregates metal on the surface of the melt, as well as reducing convection of the metal, which in the absence of noticeable solubility, is the major mechanism of yield loss in the electrolytic cell.
- Lithium is less reactive than other alkali metals toward oxygen and water, so that the surface of both solid and liquid lithium takes noticeably longer to dull when exposed to the atmosphere. This is a kinetic effect due to the comparatively high atomization and ionization energies of lithium; the heat of formation of lithium oxide and hydroxide are greater than for other alkali metals. Unlike the other alkali metals, lithium reacts directly with nitrogen and its nitride is stable at the cell temperature. Nitridation, however, is a slow process, and oxidation prevails in the cell.
- A less well-known feature, which has substantial impact on the efficiency of electrolytic production, is the great affinity of lithium for molten LiCl, likely due to the formation of a subchloride Li_2Cl layer [12], which as described in Mellor [6], manifests itself in a thin, protective LiCl coat. In the present experiment, the integrity of this layer holds even for lithium globules several centimeters across despite the fact that their apex, due to the surface tension and density differential, is raised by as much as 1 cm above the melt. The presence of this LiCl layer can be easily ascertained by scraping the surface of a molten lithium globule. Its importance for the present application lies not just in the protection it affords from the atmosphere, but more so from chlorine gas formed at the anode and transported by convection throughout the electrolyte. Thus, the hot chlorine corrodes a stainless steel rod immersed in the melt at a rate of 100–200 μm/h (forming a liquid $FeCl_3$ top layer), but it does not react with the lithium.

Figure 4.1 shows a lithium globule of 0.52 mol on the surface of the melt after electrolysis. The normally transparent thin LiCl layer is made visible by a small amount of surface active carbon impurity from the anode adhering to the interface. From the optically measured surface area of the globule we obtain an average lithium height in the melt of 10.1 mm, using a lithium density of 0.49 g cm^{-3} [9] at the cell temperature. This is in good agreement with 11.5 mm obtained from surface tension theory (Equation 2.17–Equation 2.19) for a eutectic density and surface tension of 1.85 g cm^{-3} and 130 mN m^{-1}, respectively [13].

Figure 4.2 shows the variation of the total liquefaction temperature of LiCl–KCl mixtures as a function of LiCl content, in the vicinity of the eutectic at 352°C [14]. This variation exhibits a sharp transition in the region of the eutectic, so that allowing for a variation of composition during electrolysis of 5–10%, one can reasonably expect to operate the cell around 400°C–420°C. Because the concentration of lithium decreases as electrolysis proceeds, a cell is started with an initial LiCl-to-KCl molar ratio of about 63:37, which corresponds to a weight ratio of 50% and a mp of about 400°C. The author has found that operation of the cell a few percent on the KCl-rich side of the eutectic is accompanied by a loss of viscosity in the bath as well as increased operating voltage (as one would expect from the decreased lithium concentration), leading to increased convection as well as frequent small explosions due to recombination of the electrolysis products. Hence, fresh LiCl needs to be added to the bath after about 5% of the initial lithium content has been reduced.

FIGURE 4.1 *A color version of this figure follows page 112.* A lithium globule on the surface of a molten KCl–LiCl eutectic at 420°C after electrolysis. At this temperature, lithium ordinarily burns in air; however, a thin film of electrolyte protects it. The film is made visible by traces of carbon from the anode adhering to the surface.

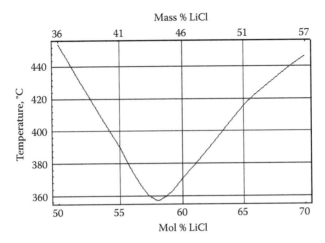

FIGURE 4.2 Complete liquefaction temperature for LiCl–KCl mixtures as a function of LiCl molar and mass fraction. The best starting point for the cell is a few percent on the LiCl-rich side of the eutectic through which the electrolyte passes as the electrolysis proceeds.

Graphite, being 20–30% porous [15], is not impermeable to the LiCl–KCl electrolyte, which slowly diffuses toward the outer walls of the cell until saturation is reached. This diffusion is very slow (<100 μm/h) when the cell passes no current; however, it accelerates markedly during electrolysis so that saturation at the outer walls for 12.5 cm wall thickness is reached in about 4–5 h. This phenomenon is due to the finite conductivity and, hence, the voltage drop of the graphite cell walls; the latter is decreased by diffusion of the electrolyte into the graphite pores, thus minimizing the electric field energy. If the electrolyte saturates the contact area where the metal strap forms an electrical connection to the graphite, the contact is rapidly oxidized, and the circuit is subsequently closed by the electrolyte, resulting in the metal strap acting as a surrogate anode, with an increased cell voltage. The evolution of Cl_2 is then markedly reduced, and the anodal process is replaced by the oxidation of iron:

$$2Fe + 3Cl_2 \rightarrow 2FeCl_3 \; \varepsilon_{420} = 0.94 \text{ V} \qquad (4.1)$$

This effect can be minimized by locating the strap at the upper lip of the crucible substantially above the level of the electrolyte in the melt so that diffusion is opposed by gravity, while by arranging the topmost part of the crucible to protrude from the oven, the temperature of the contact region can be maintained below the mp of the melt, preventing the salt from reaching the contact area. Alternatively, electrical contact can be achieved by means of several graphite rods pressed against the top lip of the crucible.

4.3 EXPERIMENTAL

The properties of lithium described earlier make stringent temperature control and cathode cooling unnecessary in the lithium cell.

A graphite crucible with 63 mm (2.5 in.) internal diameter and 7–10 cm deep, with a cylindrical stainless steel contact strap about 2 cm wide attached around its lip, is placed on a porcelain plate inside a top-loading crucible oven. An LiCl–KCl mixture of equal proportions by weight is added to the crucible until the electrolyte is about 5–6 cm deep when molten, the crucible thus holding about 3.5 mol LiCl prior to electrolysis. The temperature of the oven is raised so that a thermocouple placed in the melt indicates a melt temperature of about 410°C–420°C; the oven temperature is then maintained at the corresponding value (about 440°C) by electronic control. The time–temperature curve for the melt should show an inflection at around 355°C corresponding to melting of the eutectic, followed by a region of decreased inclination as the mixture gradually liquefies. The rate of temperature rise should increase again at about 410°C, corresponding to complete melting, and indicating the mixture is of correct proportions.

The thermocouple is removed once thermal equilibrium is reached to avoid corrosion by chlorine, and cell temperature is then controlled by keeping a constant oven temperature. A voltage of about 6.7 V is applied to the cell, and a steel electrode of about 3 mm diameter is lowered until it just makes contact with the melt. The current should rise to 4–5 A within 30 sec.

The use of a graphite rather than a porcelain cell eliminates the need for careful drying to avoid LiOH formation; however, a freshly fused mixture normally contains a substantial amount of water due to the deliquescent nature of LiCl (it completely dehydrates only slowly above 186°C [16]). Electrolyzing the mixture at this stage leads to hydrogen evolution at the cathode at a low current because the cathode resistance is high with no lithium present. Holding the fresh electrolyte for several hours at 420°C is normally sufficient to dehydrate the cell enough for lithium reduction to begin immediately at the cathode; alternatively, the cathode can be lowered into the electrolyte to raise the current until cathodic gas evolution ceases.

As lithium starts forming, some red sparking due to convective diffusion to the anode can initially arise; however, after about 0.02 mol lithium has formed, the cathode is raised so that contact with the melt occurs only through the lithium already formed. This reduces convective diffusion due to the surface tension of the lithium, and the increasingly growing globule is prevented from detaching from the cathode by the high surface tension of the latter, as well as by the electric field in the melt (as described in Chapter 2).

As the lithium globule diameter increases to a few centimeters, the cathodic resistance and concentration over-voltage decrease, so that the current rises. The latter is clamped by computer control to about 22 A, to avoid excessive convection and effervescence, which gradually decreases the cell voltage to about 6.5 V. In addition, with increasing globule size the effectiveness of surface tension in keeping a minimum perimeter decreases, while the electric field attaching it to the cathode produces a force that stretches it toward the anode. Contact with the anode is prevented by a proportionally greater chlorine evolution at the point of closest approach. The lithium does not react with Cl_2 effervescence transported through the electrolyte by convection currents, due to the LiCl layer described earlier. This situation is shown in Figure 4.3. Electrolysis is discontinued when either a charge of about 0.7 mol electrons has passed through the cell, at which point the electrolyte is 5% KCl rich

FIGURE 4.3 *A color version of this figure follows page 112.* The molten lithium acts as the cathode being attracted to the steel electrode by electrical forces. This arrangement uses the surface tension of lithium to reduce its fragmentation, which otherwise leads to small lithium globules reaching the anode where their protective electrolyte coat is broken and an explosive reaction with chlorine results.

FIGURE 4.4 *A color version of this figure follows page 112.* Lithium is collected from the cell with a perforated ladle. With the protective coat of electrolyte broken, surface oxidation commences and the molten lithium may ignite. It is thus quickly run into paraffin and held down with steel gauze.

with respect to the eutectic, or the circuit is broken due to the globule detaching from the cathode (this does not occur earlier than about 0.4 mol). The lithium, which now looks as in Figure 4.4, is scooped out with a semicircular perforated (1 mm holes) ladle (it is very difficult to remove large globules with a flat spoon as suggested in Reference 6), and the metal is dipped in paraffin. Very little electrolyte adheres to the metal, and any present can be easily separated with a blade.

REFERENCES

1. Dönges, E., Alkali metals. In *Handbook of Preparative Inorganic Chemistry, 2nd ed.,* p. 957, edited by Brauer, G. New York–London: Academic Press, 1963.
2. Furniss, B. S., Hannaford, A. J., Smith, P. W. G., and Tatchell, A. R., *Vogel's Textbook of Practical Organic Chemistry, 5th ed.,* pp. 1070–71. London: Addison Wesley Longman, 1989.
3. Sandler, S. R., and Karo, W., *Organic Functional Group Preparations*, p. 58. New York–London: Academic Press, 1968.
4. Dillich, S., Hydrogen Storage Program Element Introduction. In 2009 Annual Progress Report, November 2009, published on DOE/EERE Program Web site. http://www1.eere.energy.gov/hydrogenandfuelcells/storage/metal_hydrides.html
5. Roscoe, H. E. and Schorlemmer, C., *A Treatise on Chemistry,* Vol. II, pp. 157–161. London: Macmillan & Co., 1879.
6. Mellor, J. W., *A Comprehensive Treatise on Inorganic and Theoretical Chemistry,* Vol. 2, p. 449. London–New York: Longmans Green, 1927.
7. Meyer, H. C., Some practical aspects of handling lithium metal. *Hand. Uses Alkali Metals, Adv. Chem.* 19: 9–15, 1957.
8. Guntz, M., Sur la Préparation du Lithium Métallique. *Compt. Rend.* 117: 732, 1893.
9. Kipouros, G. J. and Sadoway, D. R., Toward new technologies for the production of lithium. *JOM,* 50(5): 24–26, 1998.
10. Ruff, O. and Johannsen, O., Über die Gewinnung von Metallischem Lithium. *Zeit. Elektrochem.* 12: 186, 1906.
11. Baker, P. S., Wells, G. F., and Rathkamp, W. R., A cell for the preparation of small quantities of alkali metals. *J. Chem. Educ.* 31(10): 515, 1954.
12. Hébant, P. and Picard, G. S., Electrochemical investigations of the liquid lithium/(LiCl-KCl Eutectic Melt) interface. Chronopotentiometric and electrochemical impedance spectroscopy measurements. *Electrochim. Acta* 43(14–15): 2071–81, 1998.
13. Nissen, D. A. and Carlsten, R. W., Surface tension of LiCl-KCl eutectic mixture. *J. Chem. Eng. Data* 18(1): 75–6, 1973.
14. Basin, A. S., Kaplun, A. B., Meshalkin, A. B., and Uvarov, N. F., The LiCl-KCl binary system. *Russ. J. Inorg. Chem.* 53(9): 1509–11, 2008.
15. Hupp, T. R. et al., Graphite, Artificial. In *Kirk-Othmer Encyclopedia of Chemical Technology, 5th ed.* Vol. 12. New York: John Wiley & Sons, 2006.
16. Masset, P. J., Thermogravimetric study of the dehydration reaction of LiCl·H_2O. *J. Therm. Anal. Calor.* 96(2): 439–41, 2009.

5 Cesium

SUMMARY

- CsCl is thermally reduced by calcium at 640°C–700°C in the vacuum of a single-stage pump.
- Quartz is attacked by calcium vapor, but not by cesium vapor below 700°C.
- The cesium yield is ~6 g in a single run, or 40–48% based on CsCl; unreacted CsCl is recovered.
- Cesium purity is >98% by IC analysis, main impurity 1% rubidium; no calcium evident.
- Availability and cost of CsCl compared to cesium metal make method economical

APPLICATIONS

- Lowest melting alkali eutectic with 24% K and 3% Na (mp −72°C) [1].
- Getter material with much lower O_2 partial pressure than Na, K, or Ba [1].
- Preparation of cesium organometallics offering exceptional reactivity [1,2].

5.1 INTRODUCTION

Cesium salts are readily available and only moderately expensive. However, elemental cesium, which is the starting point for the foregoing applications, is a much rarer commodity and far more expensive. Hence, preparation in the laboratory is an attractive proposition both economically and timewise.

Cesium and rubidium are rare alkali metals with specialized uses in physical optics and electronics, which have recently also found use in organic synthesis. In physics applications, the photo-emission/absorption spectrum of rubidium atoms makes its vapor useful in studying nonlinear photonic and soliton interactions [3]. Cesium is one of three metals (the others being mercury and gallium) that is liquid near room temperature. Its low ionization potential makes cesium useful as a photocathode in photoemissive devices [1]. As a chemical reagent, cesium forms alkyl and aryl organometallics of exceptional reactivity, which can be effective alkylating agents when organometallics of other alkali metals fail [1,2].

Cesium and rubidium were first discovered in their salts by Bunsen and Kirchoff in 1860 [4], following Kirchoff's invention of the emission spectrophotometer using the hot nonluminous flame of the Bunsen burner. Cesium manifested itself in the spectrum of naturally occurring mixed chloride salts by the presence of two lines at the blue end of the emission spectrum. These intensified relative to the rest of the spectrum with repeated precipitation of the chloroplatinate salt (which removes the more soluble chloroplatinates of the lower alkali metals), followed by decomposition and solvation. The following year, Bunsen obtained elemental rubidium by the thermal decomposition of the tartarate salt and measured its melting point [5].

Elemental cesium was first prepared by electrolysis of its molten cyanide by Setterberg in 1882 [6]; however, due to its extreme affinity for oxygen, electrolysis had to be conducted under the strict exclusion of the latter. This, combined with the difficulty of electrolyzing small amounts of material efficiently, makes thermal reduction in a vacuum a much more attractive proposition for preparing cesium in the laboratory.

Thermal reduction of alkali salts with calcium, which is applicable to all alkali metals starting with sodium and higher, was published by Hackspill in 1905 and refined by him over several years [7–9]. He used an ordinary soda glass reactor tube carrying a tightly fitting steel liner filled with a reaction mixture consisting of the alkali salt and fine calcium turnings. The glass tube had an upper sidearm that connected to a horizontal tube carrying several glass ampoules for receiving and redistilling the product metal, and was evacuated to a medium vacuum (1 mtorr) by a mercury pump. The reactor was heated in a tube oven initially to 300°C, where gas evolution led to a temporary deterioration of vacuum, and ultimately to 700°C over 3–4 h (Hackspill reported that the yield decreases with increasing rate of heating), with the glass softening at around 650°C and hugging the liner, but maintaining the vacuum. After all the cesium was collected and heating discontinued, the reactor tube shattered due to its lower coefficient of thermal expansion. The cesium yield based on CsCl using a fourfold excess of fine calcium turnings was reported to be almost quantitative. Although the reaction did not complete until 700°C, Hackspill noted that redistillation of the alkali metals completed at a much lower temperature (250°C–300°C for potassium and higher alkali metals at 1 mtorr).

The present preparation is a modified version of the Hackspill process, making it accessible with readily available glassware, and addresses some of its drawbacks, such as the nonreusability of the apparatus. It consists of the thermal reduction of cesium chloride by coarse calcium turnings in the moderate vacuum (30 mtorr) achievable by a single-stage mechanical vacuum pump, using a quartz test tube with a ground quickfit joint. Use of quartz is allowable because it has been observed that, with the exception of lithium, alkali metals exert no effect on fused quartz in the gaseous phase and at reduced pressure up to 720°C, and in the condensed phase below 250°C.

Because the reaction is mixed-phase, the yield is very much dependent on the fineness of the calcium turnings, and though not stated by Hackspill, his choice of a fourfold excess of calcium is determined more by the surface area of the calcium turnings he used than by the actual amount of calcium present, which would have been the case if the yield was dictated by thermodynamical considerations. The author has found that even coarse turnings (0.5 mm) yield about 40–45% cesium based on CsCl; moreover, the overall loss of cesium is only very slight, about 0.4 g per run, corresponding to the metallic coat on the test tube wall, since the unreacted CsCl is easily recovered by working up the reaction products.

The method has the advantage that small impurities of lithium, sodium, and to a lesser extent of potassium, are separated from the cesium in the course of the reaction, so that even without redistillation, the product metal contains less impurities than the reagent salt. Thus, if discernible amounts of sodium or potassium are present in CsCl, they can be seen as individual shiny metallic bands deposited higher on the quartz test tube than cesium, where the tube wall temperature corresponds to their respective boiling points. Lithium does not volatilize if proper reaction temperatures are used (if it is volatilized, its vapor reduces the silicon in quartz instantly). Cesium (and rubidium), having the lowest boiling point of all alkali and alkaline earth metals, is deposited as a golden metallic band on the lowest part of the test tube, and collects in the receiver.

5.2   DISCUSSION

Cesium is thermochemically reduced by calcium in the following reaction:

$$2CsCl + Ca \rightarrow CaCl_2 + 2Cs. \qquad \begin{array}{l} \Delta H^0_{298} = +90kJ\ mol^{-1} \\ \Delta S^0_{298} = +31J\ mol^{-1}K^{-1} \\ \Delta G^0_{298} = +81kJ\ mol^{-1} \end{array} \qquad (5.1)$$

We see that the free energy change is large and positive, corresponding to a very small equilibrium constant $K_{298} \sim 10^{-14}$, so that under standard conditions this reaction will not proceed. In the present experiment, two circumstances change this.

First, the experiment is conducted at 610°C–700°C, making use of the fact that for an endothermic reaction the equilibrium constant increases with temperature. Further, since the entropy change is positive, we expect the free energy to decrease with temperature. This is indeed so, and at the boiling point of cesium at 1 atmosphere (671°C), using $\beta \equiv (\Delta H^0 - \Delta G)/T = 42.5$ kJ mol^{-1} T^{-1} at 900 K [10], we find

$$2CsCl(l) + Ca(s) \rightarrow CaCl_2(l) + 2Cs(l), \qquad \Delta G^0_{298} = 50kJ\ mol^{-1}, \qquad (5.2)$$

which is valid for the pure substances, and hence neglects the free energy of solvation (interaction) of $CaCl_2$ in CsCl. This omission shows up in the fact that $CaCl_2$, although liquid under the reaction conditions, is solid as a pure substance at that temperature. However, it is easily checked that most of the decrease in free energy with respect to standard conditions above is due to a difference in the heat capacities of the reactants and products, with the phase change of $CaCl_2$ and CsCl contributing relatively little to free energy change because fusion close to the melting point is a reversible process. Hence, for thermal reasons, the equilibrium constant at 944 K is 10 orders of magnitude greater than at room temperature $K_{944} \sim 1.7 \times 10^{-3}$.

Second, the reaction is performed at a pressure of about 30 mtorr, which is below the vapor pressure of cesium at the reaction temperature, so that the Reaction 5.2 is accompanied by a phase change:

$$Cs(l) \rightarrow Cs(g) \qquad \Delta_V H = 69kJ\ mol^{-1}. \qquad (5.3)$$

At atmospheric pressure, cesium boils at 944 K and, at this temperature and pressure, the phase change contributes nothing to the free energy because the reaction is reversible. However, at a cesium pressure of 30 mtorr, the equilibrium constant for the condensed phase is large

$$K \equiv \frac{CaCl_2}{[CsCl]^2} = 6.2 \ 10^8 K_{944} \sim 10^6,$$ (5.4)

and the reaction can proceed.

Using the latent heat of vaporization, we can calculate the variation of K with temperature around 944 K because, for small temperature changes, the contribution of the phase transition latent heat to the free energy is much greater than that of the heat capacity difference between reactants and products. Thus, the equilibrium constant is expected to vary as

$$K(T) \sim \exp\left(-\frac{\Delta H_0}{RT} + \frac{\beta}{R}\right) \exp\left(-2\Delta_v H\left(\frac{1}{RT} - \frac{1}{RT_b}\right)\right).$$ (5.5)

where the first and second factors are contributions from Processes 5.2 and 5.3, respectively. Solving Equation 5.5 for the temperature T_0 at which the equilibrium constant for the condensed phase at cesium partial pressure p equals unity, we find the *equilibrium temperature*:

$$T_0 = \frac{\Delta H_0/2 + \Delta_v H}{-R \ln p + \beta/2 + \Delta_v H/T_b},$$ (5.6)

so that at 30 mtorr, $T_0 = 640$ K. Although no signs of reaction are evident at this point when pure CsCl is used, when the melting point is lowered by the addition of LiCl, as will be explained below, substantial cesium deposition indeed commences at this stage.

Similarly, we can calculate the temperatures corresponding to unity equilibrium constant for the other alkali metals:

$$2LiCl(l) + Ca(s) \rightarrow CaCl_2(l) + 2Li(g) \qquad T_0 = 842K \qquad (5.7)$$

$$2NaCl(l) + Ca(s) \rightarrow CaCl_2(l) + 2Na(g) \qquad T_0 = 592K \qquad (5.8)$$

$$2KCl(l) + Ca(s) \rightarrow CaCl_2(l) + 2K(g) \qquad T_0 = 642K \qquad (5.9)$$

$$2RbCl(l) + Ca(s) \rightarrow CaCl_2(l) + 2Rb(g) \qquad T_0 = 622K \qquad (5.10)$$

$$Ca(s) \rightarrow Ca(g). \qquad T_0 = 900K. \qquad (5.11)$$

The equilibrium temperatures for lithium and sodium are essentially determined by the point at which their vapor pressure equals 30 mtorr, while for the remaining alkalis T_0 is higher due to the substantial lattice energy of their halides. Hackspill mentions that the thermal reduction does not work for lithium, and this is reflected in the closeness of its T_0, Equation 5.7, to the boiling point of calcium, Equation 5.11. For the other alkali metals, the reaction can be made to proceed over a temperature interval of about 250°C, with the yield increasing somewhat with temperature. It is absolutely essential, however, that the calcium should not be allowed to boil, as unlike the alkali metals (except lithium) its low-pressure vapor attacks quartz instantly, depositing a black silicon layer on its surface.

Aluminum and magnesium, which are frequently used in thermal reductions, cannot reduce CsCl due to the volatility of the higher AlX_3 halides, and the substantial vapor pressure of magnesium under the conditions of the present experiment. Reduction of the alkali fluorides is possible with aluminum, since AlF_3 is ionic with a high melting point (1260°C subl.), and aluminum has a much higher boiling point than calcium. However, its low melting point leads to stratification of the reagents in the absence of external agitation, while the maximum theoretical yield is 50% due to the formation of cesium cryolite [11]:

$$6CsF + Al \rightarrow 3Cs + Cs_3AlF_6, \tag{5.12}$$

which is not reduced further. Furthermore, it is more difficult to recover unreacted CsF than CsCl from the calcium reduction products, due to the reduced solubility of the cryolite and the inconvenience posed by HF.

Because the vapor pressure difference between the alkali metals and calcium are the driving force of the reduction in Equation 5.7–Equation 5.10, it is useful to consider the variation of metal vapor pressures with temperature with greater accuracy than provided by Equation 5.5, and this is presented in Figure 5.1 [10]. It is apparent that useful results can be obtained in a temperature range restricted from T_0 to near the bp of the reducing agent. For this reason, good temperature control is required, and even more importantly, the reagents must be kept in a uniform temperature environment throughout the reaction to prevent vaporization of calcium in high-temperature regions prior to completion of the reduction in the low-temperature regions. To achieve a more uniform temperature control, a box oven is used here rather than the tube oven of the Hackspill experiment.

As was noted by Hackspill, the cesium yield is strongly dependent on the heating rate, and this combined with the fact that reaction temperatures are considerably higher than those predicted by thermodynamics, strongly suggests that the temperature at which reaction is observed to occur is determined by the melting point of the reagents. The starting temperature corresponds to the point at which the CsCl melts and wets the surface of the calcium turnings, with the reaction taking place on this interface. This is supported here by the observation that no cesium is distilled prior to about 640°C if very pure CsCl is used (the mp for pure CsCl is 646°C), while with technical-grade CsCl containing a few percent NaCl and KCl impurity, distillation commences at about 610°C and is essentially complete by 640°C. Furthermore, the

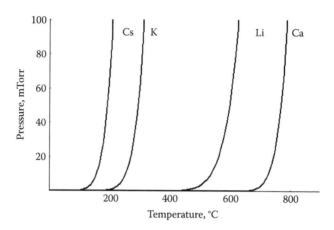

FIGURE 5.1   The temperature variation of the saturated vapor pressures of the alkali metals and the calcium metal reductant showing the temperature range over which the reaction can be carried out.

thermal reduction of a CsCl/LiCl eutectic consisting, of a 40/60 molar ratio with a melting point of 315°C, is already well under way at 400°C.

The reactions proceed as follows: As the CsCl reacts to form dissolved $CaCl_2$, the melting point of the local salt layer rises, since the mp of pure $CaCl_2$ is 772°C. To prevent the reaction being halted by the early formation of a solid $CaCl_2$ layer, the reaction rate must be slow compared to the time required for the $CaCl_2$ to equilibrate in the reagent mixture. Hence, a slower temperature rise produces higher yields as Hackspill observed. Because the mp of $CaCl_2$ is well above the bp of calcium in equilibrium with its vapor at 1–10 mtorr, and this vaporization must be avoided to prevent calcium vapor reacting with the quartz and contaminating the product, it is clear that complete conversion is impossible. Because the fused salts are denser than calcium turnings, as the salt melts it percolates through the calcium turnings until they eventually float on the surface of the melt. From this, it is also clear that it is desirable to maximize the surface area of the molten salt pool, for which purpose the reaction tube is tilted only slightly from the horizontal, sufficiently to contain the reagents. Mechanical and/or ultrasonic mixing will undoubtedly increase the yield beyond that obtained here; however, external stirring of an evacuated glass tube bearing a steel insert would substantially raise the complexity of the experiment.

Because the reaction temperature is determined by the melting point of the reagents, LiCl can be added to reduce the end-point temperature from 700°C to around 550°C due to its effectiveness in lowering the mp of CsCl. This has little influence on the yield; indeed, if substantial amounts of LiCl are used, the yield drops due to the reduced CsCl concentration in the melt. However, the lower temperature makes temperature control simpler since it increases the gap between the end-point temperature and the bp of calcium. The lithium itself does not contaminate the product since its vapor pressure is very low at the new end-point temperature, and it does not alloy with cesium.

5.3 EXPERIMENTAL

As mentioned above, a quartz test tube reactor stands up well to alkali metal vapors at reduced pressure, and eliminates the softening and ultimate shattering of the soda glass tube used in the Hackspill experiments. The quartz tube is fitted with a 24/29 mm quickfit socket, and is 28 mm in diameter and 250 mm long. To avoid condensed alkali metal flowing back into the high-temperature region where it can damage the quartz, the tube is clamped at an inclination of about 10° to the horizontal (which suffices to ensures the fused salts are contained), with the open end pointing down. This arrangement minimizes loss of metal that occurs in the more common vertical arrangement when vapor condenses in the upper section of the tube. The shallow tube angle maximizes contact between the calcium turnings and the fused salt bath by increasing the surface area of the latter.

A 120-mm-long stainless steel tube with 16 mm internal diameter, and welded closed at one end, is used as the reagent liner. The liner walls are 1.6 mm thick so that its thermal resistance is much lower than that of the gap between the liner and the glass walls. This serves to improve the temperature homogeneity of the reagents, which is important due to the narrow temperature range over which the reactor operates. A hole ~1 mm deep drilled in the closed end of the liner receives the end of a stainless steel rod used to support the liner in the upper end of the quartz tube. There must be a space of about 10 mm left between the closed end of the quartz tube and the open end of the liner to allow the alkali metal vapors to escape.

The quartz tube socket is attached to a glass tube that carries a connection to the vacuum line, as well as an ampoule for receiving the product metal, welded at 90° to the tube. This glass tube is not exposed to high temperatures and can be ordinary borosilicate or soda lime glass. This, however, introduces the possibility of the joint shattering due to the larger expansion coefficient of the borosilicate compared to quartz [12]. Swapping the placement of the cone and socket creates the same problem on cooling, since atmospheric pressure will push the borosilicate socket into the quartz during heating. For this reason, the joint is cooled from the outside by several turns of water-cooled copper tubing, with the contact thermal resistance decreased by the application of some heat sink compound and spring compression applied to the copper turns.

The reagents consist of a 1:2 w/w mix of coarse calcium turnings and cesium chloride powder. In this ratio the tube described above holds about 8 g of calcium turnings and 16–20 g CsCl. The turnings need to have a clean shiny surface. They can, for instance, be freshly prepared by shearing calcium granules into slices ~ 0.5 mm thick. The CsCl need not be fused as described by Hackspill [7–9]. No observable moisture is retained if the CsCl is simply held at about 250°C for 1 h prior to use. The ultimate reaction temperature can be decreased by about 100°C if 10–20% by weight anhydrous LiCl is mixed uniformly with the CsCl to lower the melting point.

The tube containing the reagents is shaken in a horizontal plane to ensure even mixing and avoid the heavier and finer CsCl segregating at the bottom. The mixture is then hand-pressed with a suitable plunger, and a small plug of steel wool inserted to cover the opening. The liner is then inserted into the quartz tube by pushing it with the end of a steel rod held in the hole in its base. The glass tube with the vacuum inlet and ampoule is now attached to the greased quickfit joint, so as to hermetically seal the liner and supporting rod. The entire apparatus, located inside a suitably fitted box oven, is shown in Figure 5.2.

The vacuum pump is turned on, and the pressure allowed to drop below 35 mtorr. At this stage, the oven temperature is raised uniformly over about 45 min to 450°C if LiCl was added, and to 550°C otherwise. If the calcium turnings contained patches of hydroxide, some gas evolution and a vacuum deterioration (which should be no worse than 100 mtorr) will be evident at about 500°C. When the set temperature is reached, the salt starts melting, wetting the surface of the calcium granules, and reacting according to Processes 5.2 and 5.3 above. This is evidenced by a metallic mirror appearing on the walls of the quartz tube immediately above the cooling coils (Figure 5.2). If some potassium or sodium impurity is present in the reagent, these metals condense in bands spread in accordance with local wall temperature along the tube.

From this point onward, the temperature is raised slowly to compensate for the decreasing concentration of CsCl in the melt while allowing $CaCl_2$ to diffuse from the CsCl–Ca interface and be replaced by fresh CsCl. Raising the temperature too rapidly results in a loss of yield [7–9], as the unreacted CsCl collects in a molten pool at the base of the liner. A heating rate of 40°C–50°C/h up to a maximum temperature of 150°C above the base temperature has been found adequate. The temperature must not be allowed to exceed 700°C (640°C if LiCl was added), since the calcium vapor pressure at this temperature is sufficient to attack the quartz reactor strongly, reducing its surface to a black silicon crust (the boiling point of calcium at 30 mtorr is 750°C).

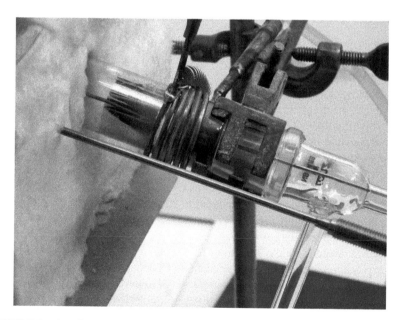

FIGURE 5.2 *A color version of this figure follows page 112.* The evacuated lower end of the quartz reactor tube with side arm and detachable cesium receptor. Cesium condenses ahead of the region cooled by the copper tubing, coalesces into drops and flows into the receptor.

As the temperature slowly rises, droplets of molten cesium start forming inside the mirrored surface of the collector tube, and flow down as they attain sufficient size, forming a pool of molten cesium. At the 10–30 mtorr pressure of a single-stage pump, a few isolated spots of higher oxides of cesium (mostly the purple Cs_9O [13]) are seen to form on the cesium surface; however, as the cesium flows into the collector and ampoule it sheds this crust, which remains stuck to the glass (Figure 5.3). Some gentle tapping, as well as slight heating with an air gun, might be required to ensure as much cesium as possible collects in the receiver (at this pressure the cesium easily flows down a narrow ampoule neck). When the final temperature is reached, it is maintained for about a half hour to ensure no more cesium flows into the collector, whereupon heat from a propane burner is applied evenly around the circumference of the ampoule neck; as the glass softens and collapses under pressure, the ampoule is gently pulled away from the apparatus and sealed. The yield of cesium over several runs varied in the range 5–6 g (about 0.5 g remains stuck to the reactor walls), representing a 40–48% yield.

Cesium can be stored under a 4–5 cm deep layer of sodium-dried paraffin oil without substantial oxidation for several months. If this is desired, after no more cesium collects in the receiver, the tube is allowed to cool under vacuum to room temperature (it can be withdrawn from the oven after the temperature has dropped below 500°C to speed cooling), whereupon the vacuum is removed, the quickfit joint separated, and the receiver containing the cesium quickly dipped below the surface of paraffin oil warmed above the melting point of cesium (28.7°C). Although cesium is generally considered pyrophoric in air, it is found that no observable oxidation occurs if the above procedure is performed quickly, while the sublimates in the form of a metallic mirror on the tube surface oxidize rapidly in air, but do not catch fire. When dipped below the surface of warm paraffin, cesium collects in a golden globule (the gold coloring is due to a trace amount of oxides; absolutely pure cesium is silver

FIGURE 5.3 The temperature gradient at the exit from the reactor tube creates a zone where traces of alkali metal impurities in the CsCl condense in accordance with their bp. Cesium has the lowest bp and collects in the receiver.

FIGURE 5.4   Cesium collected in a tube protected by liquid paraffin.

colored). Any oxides formed from the brief period of exposure to air are pushed to the surface by surface tension, where they either separate or can be skimmed from the surface (Figure 5.4).

The unreacted cesium chloride mixed with calcium and calcium chloride in the steel liner is fairly easily extracted by dissolving the contents of the liner in water, and after hydrogen evolution has ceased filtering to remove the calcium hydroxide. The remaining hydroxide in the filtrate is then neutralized with HCl, and $CsCO_3$ solution added precipitating $CaCO_3$ until the pH starts rising. The solution is then filtered and evaporated to yield CsCl.

The quartz tube can be cleaned from the adhering oxides by slowly adding *t*-butyl alcohol in drops down the tube walls, and rinsing in water when hydrogen evolution has ceased. There is often a superficial brown discoloration on parts of the quartz tube in close proximity to the mouth of the liner, due to reduction of the quartz by a small amount of calcium vapor carried by the cesium. This can be removed by brief cleaning in 4–5% HF solution (5–6 g KF dissolved in 40–50 mL conc. HCl).

REFERENCES

1. Ferguson, W. and Gorrie, D., Cesium and cesium compounds. In *Kirk-Othmer Encyclopedia of Chemical Technology, 5th ed.*, Vol. 5. New York: John Wiley & Sons, 2006.
2. Astruc, D., *Organometallic Chemistry and Catalysis*, pp. 289–312. Berlin–Heidelberg: Springer-Verlag, 2007.

3. Dawes, A. M. C., Illing, L., Clark, S. M., and Gauthier, D. J., All optical switching in rubidium vapor. *Science* 308(5722): 672–74, 2005.

4. Bunsen, R. W., Gewinnung der Rubidiumverbindungen. *Eur. J. Org. Chem. (Annalen)* 122: 347–354, 1862.

5. Bunsen, R., Über die Darstellung und die Eigenschaften des Rubidiums. *Eur. J. Org. Chem. (Annalen)* 125: 367, 1863.

6. Setterberg, C., Über die Darstellung von Rubidium- und Cäsiumverbindungen und Über die Gewinnung der Metalle selbst. *Eur. J. Org. Chem. (Annalen)* 211: 100–116, 1882.

7. Hackspill, L., Sur une Nouvelle Préparation du Rubidium et du Caesium. *Compt. Rend.* 141: 106–7, 1905.

8. Hackspill, L., Preparation des métaux alcalins. *Bull. Soc. Chim. Fr.* 9: 446–51, 1911.

9. Hackspill, L., Sur quelques propriétés des métaux alcalins. *Helv. Chim. Acta* 11(1): 1003–26, 1929.

10. The National Institute of Standards and Technology (NIST) Virtual Library http://webbook.nist.gov/chemistry/form-ser.html.

11. Chemische Fabrik Griesheim-Elektron, Verfahren zur Darstellung von Alkalimetallen. DE. Patent No. 140737, Mar. 27, 1902.

12. Schenk, P. W. and Brauer, G., Preparative methods. In *Handbook of Preparative Inorganic Chemistry, 2nd ed.*, p. 10, edited by Brauer, G. New York–London: Academic Press, 1963.

13. Greenwood, N. N. and Earnshaw, A., *Chemistry of the Elements, 2nd ed.*, p. 84. Oxford, U.K.: Butterworth-Heinemann, 1997.

6 Lithium Hydride and Sodium Hydride

SUMMARY

- Hydrogen passed over lithium at 700°C generates LiH at 0.7 mol/h in quantitative yield.
- Sodium evaporated at 600°C into static hydrogen generates ~0.3 g/h NaH, free of hydrocarbons.
- Evaporation-limited reaction rate for NaH is calculated and compared to experiment.

APPLICATIONS

- Strong base in the Claisen condensation synthesis of β-diketones [1]
- Base-catalyzed elimination in the Wittig reaction [2]
- Clean metallation agents for active hydrogen compounds [3,4]
- Reduction of alkyl halides by LiH in diethyl ether or THF [5]
- Synthesis of metastable hydrides (e.g., ZnH_2, $LiEt_3BH$) by metathesis [6]
- Descaling and reduction of metal surfaces by NaH [7]
- Hydrogen storage at high hydrogen gravimetric density

6.1 INTRODUCTION

The hydrides of lithium and sodium are more stable than those of other alkali metals and, hence, are more widely used. They are useful for their strong basic properties as described earlier (they belong to the category of so-called superbases), and to a lesser extent for their reducing properties. Their use as reducing agents in organic chemistry is limited by their low solubility in inert solvents. Their role in this regard is generally restricted to precursors of the more active and/or selective $LiAlH_4$, $NaBH_4$, $LiBHR_3$, and $AlHR_2$.

The production of alkali metal hydrides by the reaction of sodium/potassium with hydrogen was first observed by Gay Lussac in 1811 [6]; Hautefeuille and Troost in 1874 established the equilibrium nature of these reactions by performing a thermodynamic analysis of the vapor pressure of the elements over the hydrides. In 1916, Lewis [6] suggested that the alkali metal hydrides are salt-like in nature, with positively charged metal cations and anionic hydrogen H^-, and this was subsequently demonstrated for LiH by x-ray diffraction spectra.

A detailed description of the preparation of LiH and NaH in the laboratory has been provided by Zintl and Harder [8] and reported by Dönges [9]. Their equipment consisted of a horizontal quartz tube placed inside a tube oven with the quartz reactor protected by a seamless steel liner (two in the case of lithium) extending well beyond the heated region. The liner was decarbonized by soaking for many hours at 900°C in moist hydrogen. A steel boat containing freshly cut alkali metal was placed inside the liner. One end of the quartz tube was hermetically attached to a manometer, while the other joined to a horizontal vessel into which the boat could be pushed on completion of the reaction, and which contained a hydrogen inlet. The hydrogen was preliminarily dried by passage through a P_2O_5 filled U-tube. In the case of sodium, the reaction was carried out between 300°C and 400°C in stationary hydrogen at atmospheric pressure, with the hydride deposited on both sides of the tube outside the heated zone. Hydrogen was added as the pressure dropped, with the reaction terminated when no more absorption occurred over a period of 24 h. In the case of lithium, the oven is heated to 700°C, at which point the LiH product is a liquid and absorption proceeds rapidly. The LiH was recovered by pulverizing the product inside the boat by a rotary mill cutter while still under a hydrogen atmosphere.

The author has found that the formation rate of sodium hydride by union of the elements at atmospheric pressure is fairly slow: ~0.2–0.3 g/h. However, the reaction is fairly straightforward and requires little input on the part of the experimenter. The procedure produces a pure white product from which oil or metal impurities do not need to be removed, as is often the case with the commercial material. For many preparations requiring only a small amount of sodium hydride [1,2], the present synthesis is suitable.

The temperature range of 300°C–400°C chosen by Zintl and Harder is overly conservative with no discernible product produced over periods of several hours. Maintaining the oven temperature at 600°C–640°C (with the tube temperature about 40°C–60°C lower, depending on the setup) produces a much faster reaction without observable contamination of the hydride with sodium. It is inconvenient to have the hydride deposited at both ends of the apparatus; hence, a reactor tube closed at one end heated by a box oven is used here. Another drawback of the original approach is that the hydride formed eventually blocks the entrance to the reactor tube (only about 0.3–0.4 g of product are sufficient for this) halting the reaction. This does not occur for lack of hydrogen over the sodium, but rather because the experiment relies on a temperature gradient in the apparatus, with the hydride formed in the 300°C–400°C region. Once this region is occupied, net production ceases because the vapor pressure of hydrogen over the hydride at higher temperatures is greater than 1 atm (see section 6.2). To overcome this problem, a long-arm spatula is sealed into the apparatus, which enables the hydride plug to be transferred into the low temperature region of the reactor and the formation of product to continue.

While the formation rate of NaH can be substantially increased by use of an inert dispersion media at high pressures [9], the author has found that at atmospheric pressure with ordinary mechanical stirring at 250°C no substantial increase in reaction rate occurs. The preparation described by Mattson et al. [10], reporting 200 mL/min hydrogen absorption at atmospheric pressure in a 1 L flask at 280°C, therefore, is entirely dependent on the special apparatus employed there, consisting of a 10,000 rpm stirrer with a baffled antivortex flask.

In contrast to sodium, the preparation of lithium hydride in the laboratory is fast and almost quantitative. This is due to the higher boiling point of lithium as well as the greater stability of LiH compared to NaH, with the higher lattice energy of LiH determined mainly by the smaller difference in the cation and anion dimensions (LiH melts at 680°C and decomposes at about 900°C at 1 atm; NaH decomposes at about 450°C). As a result, lithium hydrogenation is carried out at an oven temperature of 730°C, which is above the melting point of the hydride, reducing the kinetic impediment. The reaction is steady, but not highly exothermic or vigorous as sometimes suggested [11]. In the experimental setup presented here, 0.5–1 mol of LiH can be produced in just over an hour with a quantitative yield. It is important not to let the reaction linger because lithium has a substantial diffusion rate through steel, and can attack the quartz reactor. As in the case of sodium, the reactor tube is inclined to increase the amount of material that can be handled by the apparatus, while maintaining a large surface area of the reagents to improve the kinetics in this heterogeneous reaction.

6.2 DISCUSSION

The alkali and alkali earth metals dissolve hydrogen exothermically (a property shared with the Ti and V subgroup of metals, the rare earths, and palladium) corresponding to the formation of a hydride phase [6]. Figure 6.1 shows the variation of hydrogen partial pressure at equilibrium over lithium and sodium as a function of mole fraction of atomic hydrogen present [6], from which we see that, for hydrogen mole fractions greater than about 0.05–0.1, this partial pressure is essentially independent of composition, varying only with temperature. Since this corresponds to only one degree of freedom in the system, using Gibb's phase rule

$$F = C - P + 2,$$ (6.1)

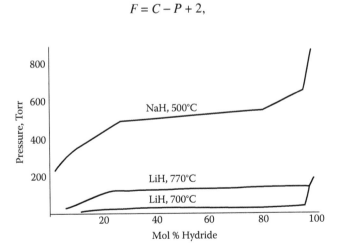

FIGURE 6.1 A variation of hydrogen partial pressure at equilibrium over lithium and sodium as a function of mole fraction of atomic hydrogen added [6]. The plateau demonstrates the presence of a hydride phase.

TABLE 6.1

Thermodynamic Parameters of the Lower Alkali Hydrides at 700 K

	ΔH_{700} kJ mol^{-1}	ΔG_{700} kJ mol^{-1}	Lattice Energy kJ mol^{-1}	mp °C
LiH	−92.7	−32.2	916	688
NaH	−56.5	2.6	791	d(420)
KH	−57.9	NA	720	NA

Source: The National Institute of Standards and Technology (NIST) Virtual Library http://webbook.nist.gov/chemistry/form-ser.html.

where C, P, and F are the number of components, phases, and degrees of freedom present in the system, respectively, we see that with the number of components $C = 2$, the number of phases in the plateau region equals 3. Because two of these phases are obviously hydrogen in the gas phase and metal saturated with hydrogen in the condensed phase, there is a strong indication for the presence of a third phase consisting of the metal hydride (saturated with alkali metal). Absence of a metal hydride phase at low hydrogen content gives $F = 2$ in that part of the graph, explaining the variation of partial pressure with composition in this region. When the mole fraction of hydrogen approaches and exceeds 1, the pressure again varies with composition, demonstrating that only two phases are now present. This indicates that the metal phase has disappeared, the alkali being entirely present in the hydride phase at 100 mol% atomic hydrogen. Thermodynamics thus indicates the formation of compounds with the formula LiH and NaH, respectively [6].

The much higher partial pressure of hydrogen over NaH than over LiH at equivalent temperatures suggests that the enthalpy of formation of LiH is substantially greater, and this is supported by the thermodynamic data presented in Table 6.1 for the representative temperature of 700 K.

Although hydrogenation is exothermic, it is not sufficiently so for sodium and potassium, so that, already at 700 K, the large negative entropy change associated with the absorption of hydrogen gas makes the free energy favor decomposition of the hydride at atmospheric pressure. Thus, although the hydride formation rate markedly increases with temperature, this has to be accompanied by the application of high pressures in the case of sodium and potassium to stabilize the hydride.

6.2.1 HYDROGENATION OF SODIUM AT ATMOSPHERIC PRESSURE

Sodium hydride in the present experiment is formed by the free evaporation of sodium from the mouth of a long cylindrical steel tube held at a fixed temperature T, substantially below the boiling point of sodium, into stationary hydrogen at 1 atm inside a quartz reactor (Figure 6.2). The walls of the steel tube are sufficiently thick to provide for a thermal conductance much greater than that of the gap around the tube, so that we can assume it to be of uniform temperature. This temperature determines the rate of sodium evaporation, and is controlled so that it does not exceed the rate

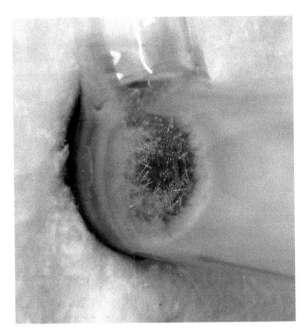

FIGURE 6.2 *A color version of this figure follows page 112.* Sodium evaporating from an oven at 600°C, in hydrogen at 1 atm. NaH needles form in a section of the reactor where the temperature is such that hydrogen pressure above NaH is below 1 atm.

of hydride formation, otherwise sodium is deposited outside the tube in the form of a metallic mirror on the quartz walls, with globules contaminating the hydride salt.

The rate of sodium evaporation can be approximated by the rate of diffusion of a gas at pressure p through a temperature gradient established in the reactor between the steel tube opening of area A at temperature T, and the temperature at the hydride plug T_0, separated by a distance L from the tube mouth in the x direction. T_0 can be estimated from Figure 6.1 as the temperature at which the hydrogen pressure over NaH equals one atmosphere; however, the evaporation rate is fairly insensitive to its exact value.

Fick's law of diffusion gives $J = -D\, dp/dx$, where J is the pressure flux of sodium vapor producing an increase in the amount of sodium outside the steel tube. Applying the ideal gas law gives the molar current I of sodium out of the tube:

$$I = \frac{JAL}{RT} = -\frac{DAL}{RT}\frac{dp}{dx},$$ (6.2)

where we have used the fact that all sodium particles are converted into NaH in volume $V = AL$ between the tube mouth and the hydride plug, with no sodium flux out of this volume. The pressure gradient dp/dx is determined by the temperature gradient at the mouth of the tube. Assuming equilibrium sodium vapor pressure at this point, and neglecting the residual sodium pressure at the hydride plug, the pressure differential equals the pressure of sodium vapor at the tube mouth at temperature T, which by reference to Chapter 5, Equation 5.5 is

$$\Delta p = p_0 \exp\left(-\Delta_v H\left(\frac{1}{RT} - \frac{1}{RT_b}\right)\right),$$

(6.3)

where p_0 is atmospheric pressure. Combining Equation 6.2 and Equation 6.3 and neglecting the small volume contraction of sodium vapor between the tube mouth and the NaH plug, we obtain the *sodium flux limited reaction rate*

$$I \sim \frac{DAp_0}{RT}\exp\left(-\frac{\Delta_v H}{R}\left(\frac{1}{T} - \frac{1}{T_b}\right)\right).$$

(6.4)

The diffusion coefficient D for particles in a gas can be computed as the product of the mean free path and the average speed of the particles. The Chapman–Enskog theory for a two-component gaseous mixture gives a more accurate value [12],

$$D_{AB} = \frac{0.0018583}{p\sigma_{AB}^2\Omega_{AB}}\sqrt{T^3\left(\frac{1}{M_A} + \frac{1}{M_B}\right)},$$

(6.5)

where

$$\sigma_{AB} = \frac{1}{2}(\sigma_A + \sigma_B),$$

(6.6)

where D is the diffusion coefficient in $cm^2 \ s^{-1}$, p is the pressure in atm, T is the absolute temperature in degrees Kelvin, M is the molecular weight in $g \ mol^{-1}$, σ_{AB} is the average Lennard–Jones molecular diameter measured in angstroms, and Ω_{AB} is a dimensionless function of the temperature of order 1, which can be neglected in an order of magnitude estimate.

Using Equation 6.4 and Equation 6.5 with $\sigma_A \sim 6.8$ Å being twice the sum of the covalent and atomic radii for gaseous sodium [13], we find the diffusion-controlled reaction rate for the sodium/hydrogen mixture to be 0.012, 0.03, and 0.08 mol/h at 500°C, 550°C, and 600°C, respectively. This compares to a reaction rate of 0.008–0.016 mol/h measured in the experiment at about 550°C tube temperature. The theory gives a reasonable estimate in view of the approximations made. At tube temperatures above ~600°C, the evaporation rate exceeds the kinetic rate of hydride formation as evidenced by deposition of sodium outside the reactor tube, and this forms the limiting temperature.

6.3 EXPERIMENTAL

6.3.1 SODIUM HYDRIDE

A quartz test tube about 30 mm in diameter and at least 200 mm long is placed inside a sealed tube oven or a circular opening in a box oven, and inclined upwards at 5°–8° to the horizontal. About 3–4 g of sodium metal, which has been shaken under ether to remove the last traces of paraffin, is placed inside a stainless steel tube, closed at one end, and inserted into the quartz reactor. A measured amount of steel wool at the bottom of the quartz test tube is used to adjust the location of the steel tube opening so it lies just inside the oven at the point where a temperature gradient is expected to commence. A side arm on the quartz tube leads to a U-tube containing several sections filled with P_2O_5 and separated by glass wool. The other end of the U-tube is connected to a hydrogen cylinder through a flow-rate adjusting valve. A long-arm stainless steel spatula (which can be fashioned out of a thin-diameter S/S pipe) is inserted through a seal so that its end lies just outside the zone where NaH is expected to be formed. The spatula runs through a fairly long section of straight tube prior to reaching the active zone in order to minimize the angular movement of the spatula and disturbance to the seal when removing the product. The outlet gases from the reactor are passed through an empty washbottle and an oil bubbler, which both isolates the reactor and serves as an indicator of the pressure inside.

The air inside the reactor is purged by opening the hydrogen valve until no oxygen is evident in the outlet gas, and the oven temperature is raised in the range 610°C–640°C (corresponding to a tube temperature of about 550°C–580°C). Hydrogen absorption commences at about 570°C as evidenced by a slow rise of oil inside the washbottle capillary, and the hydrogen valve is opened so that the level in the capillary remains about constant. There is some nonuniformity in hydrogen absorption with time, and the hydrogen feed rate should be adjusted on the high-side, which leads to some loss of hydrogen. Alternatively, a hydrogen balloon can be used.

The sodium hydride starts forming immediately outside the tube opening where the local temperature is below its decomposition temperature at 1 atm hydrogen pressure. Figure 6.2 shows the result after about 20 min of operation. After about 2 h, a wool-like plug of sodium hydride needles completely occupies the temperature region suitable for hydride formation and hydrogen absorption slows. The hydrogen flow rate can be increased at this point to produce positive pressure inside the reactor, and the spatula can be inserted into the active region and rotated to remove the plug into the low-temperature region of the reactor where the hydride is stable. This operation is repeated every few hours, resulting in an NaH formation rate of about 0.2–0.3 g/h. When sufficient NaH has formed, the quartz tube is withdrawn from the reactor and allowed to cool to near room temperature. At that point, the hermiticity of the apparatus can be broken, and the sodium hydride removed in the open atmosphere. The author has found that no spontaneously ignitable sublimates form in the reaction.

Raising the reactor temperature above about 640°C does not lead to an increased rate of hydride formation; rather, a gray color appears in the product corresponding to condensed unreacted sodium. Higher temperatures still lead to decomposition of the hydride already formed and its reformation in the section of the reactor, which now has the appropriate temperature. However, the higher evaporation rate also leads to sodium globules forming in the NaH matrix as well as condensation of liquid sodium on the walls of the quartz reactor. This is deleterious to the quartz tube because of the danger of liquid sodium flowing into the high temperature region and reducing the quartz in depth. Gaseous sodium on the other hand does not seriously attack quartz, producing just a superficial discoloration, which disappears (due to the silicon being oxidized back to silica) on exposure to air.

6.3.2 LITHIUM HYDRIDE

Lithium reacts much more rapidly with hydrogen because, as seen in Figure 6.1, its hydride is far more stable. A reaction temperature exceeding the melting point (mp) of the hydride (mp 683°C) can be used, and since LiH is denser than lithium metal (sp gr [specific gravity] 0.71 for LiH versus 0.49 for Li), the latter floats to the surface ensuring continuation of reaction. Using the same apparatus as for NaH production, and an oven temperature of 730°C (about 700°C tube temperature) completes the reaction in just over an hour, and produces about 0.7 mol of hydride.

In contrast to sodium, lithium diffuses to an extent through steel and, thus, a second stainless steel liner needs to be used [8]. To minimize the amount of contamination, the lithium is freshly cut and washed with ether, and 4.7 g is placed inside a 16-mm diameter steel tube, 120 mm long, and sealed at one end. The lithium may need to be tapped down lightly to fit in the tube. This tube is placed inside another steel tube in an oven, so that the opening of the tubes are well inside the oven. The reactor needs to be inclined about 10° to the horizontal to ensure containment of the molten lithium, while providing maximum surface area for this heterogeneous reaction.

The oven is heated to 730°C, with hydrogen absorption commencing at about 470°C. While the reaction is reported to be highly exothermic [11], there are no manifestations of excessive vigor here, the absorption proceeds uniformly, and there is no spattering. When the hydrogen feed rate has been adjusted for a uniform oil level in the capillary, it often does not have to be readjusted until the end of the reaction. After about 1–1.2 h, hydrogen absorption ceases abruptly. At this stage, the oven temperature can be raised 30°C or so to ensure no further hydrogen is absorbed. This will be the case if the tube temperature had been above the melting point of the LiH so that all the lithium had reacted. In the opposite case, some absorption will recur.

After the reactor had cooled to room temperature, the hermiticity can be broken and the tubes withdrawn in the open atmosphere. There is no danger of ignition of the contents at this point. The lithium hydride forms a hard transparent mass at the bottom of the inner tube and needs to be removed using a closely fitting, clean, mill cutter or shallow drill bit reaching to the bottom. After first pouring about 10 mL of THF into the inner tube to protect the LiH from the atmosphere, the tube is clamped surrounded by a small collection tray, and the LiH pulverized by applying steady pressure to the cutter. The bottom few centimeters of the hydride are very hard and one needs to ensure the cutter has reached the bottom of the tube to prevent loss of product.

From an initial charge of 4.7 g Li, 5.4 g LiH is obtained corresponding to a quantitative yield. Treating the product with water evolves the theoretical amount of hydrogen indicating no measurable amount of lithium in the product. The latter, nonetheless, had a slight gray coloring due to the nitride coat of the reagent.

Note: *LiH is slightly soluble in THF and drying the latter after the mechanical processing above produces a very fine powder that is easily dispersed in the air and is not readily visible. Even small amounts cause irritation of the respiratory tract, often after a considerable delay. It is advisable to conduct all operations in a hood with a strong draught and remove all traces of hydride on the apparatus upon completion of the procedure.*

REFERENCES

1. Tietze, L. F. and Eicher, T., *Reactions and Syntheses in the Organic Chemistry Laboratory*, Section 3.1.4. Sausalito, CA: University Science Books, 1989.
2. Furniss, B. S., Hannaford, A. J., Smith, P. W. G., and Tatchell, A. R., *Vogel's Textbook of Practical Organic Chemistry, 5th ed.,* pp. 495–500. London: Addison Wesley Longman, 1989.
3. Caubère, P., Complex reducing agents (CRAs)—versatile, novel ways of using sodium hydride in organic synthesis. *Angew. Chem. Int. Ed. Engl.* 22: 599–613, 1983.
4. Klusener, P. A. A., Brandsma, L., Verkruijsse, H. D., von Ragué Schleyer, P., Friedl, T., and Pi, R., Superactive alkali metal hydride metallation reagents: LiH, NaH, and KH. *Angew. Chem. Int. Ed. Engl.* 25(5): 465–66, 1986.
5. Sandler, S. R. and Karo, W., *Organic Functional Group Preparations*, p. 12. New York–London: Academic Press, 1968.
6. Mueller, W. M., Blackledge, J. P., and Libowitz, G. G., *Metal Hydrides*, pp. 165–240. New York: Academic Press, 1968.
7. Evans, N. L., The sodium hydride process for descaling metals, *Anti-Corr. Meth. Mat.* 3(2): 47–51, 1993.
8. Zintl, E., and Harder, A., Über Alkalihydride. *Zeit. Phys. Chem.* 14: 265–284, 1931.
9. Dönges, E., Alkali metals. In *Handbook of Preparative Inorganic Chemistry, 2nd ed.,* edited by Brauer, G. p. 971. New York–London: Academic Press, 1963.
10. Mattson, G. W., Whaley, T. P., and Chappelow, C. C., Sodium hydride. *Inorg. Synth.* 5: 10–13, 1957.
11. Kamienski, C. W., McDonald, D. P., Stark, M. W., and Papcun, J. R., Lithium and lithium compounds. In *Kirk-Othmer Encyclopedia of Chemical Technology, 5th ed.,* Vol. 15. New York: John Wiley & Sons, 2006.
12. Cussler, E. L., *Diffusion. Mass Transfer in Fluid Systems, 2nd ed.,* pp. 119–121. Cambridge: Cambridge University Press, 1997.
13. Sutton, L. E., ed., *Table of Interatomic Distances and Configuration in Molecules and Ions*. London: Chemical Society, 1965.

7 Bromine

SUMMARY

- Bromine is prepared by the reduction of 1-bromo-3-chloro-5,5-dimethylhydantoin (BCDMH) in an aqueous solution by sodium metabisulfite.
- BCDMH is reduced almost exclusively to bromine ($Cl_2 < 5\%$) at a quantitative yield.

APPLICATIONS

- Formation of organic bromides—often more reactive than the chloride equivalents [1].
- Hell–Volhard–Zelinsky bromination of carboxylic acids [2].

7.1 INTRODUCTION

Organic bromides are frequently more reactive than the corresponding chlorides due to the lower bond energy of the Br-C bond [2] and, therefore, are frequently preferred. The simplest method for preparing the bromides is by bromination; however, bromine is corrosive, toxic, and difficult to contain, so that its procurement can cause difficulties and its preparation is sometimes required. Textbook methods for the laboratory preparation of bromine include the oxidation of bromide salts by manganese dioxide or chlorine as well as the co-proportionation of bromide and bromate in solution. Strong oxidizing agents are unsuitable for oxidizing the bromide ion since the oxidation tends to progress further to bromate, entailing loss of product. The use of chlorine also is undesirable because bromine in the presence of chlorine is exceedingly volatile due to the rapid formation of BrCl gas (bp 5°C), with an equilibrium constant of about 8 in favor of BrCl under standard conditions. A further problem occurs when one attempts to separate or purify the bromine from the reaction mixture by distillation (bp 58°C), due to the formation of a solid bromine hydrate clathrate in the condenser.

The present experiment is a convenient preparation of bromine by the reduction of 1-bromo-3-chloro-5,5-dimethylhydantoin (BCDMH) in aqueous solution by sodium metabisulfite/sulfur dioxide. This halogenated hydantoin is readily available and inexpensive due to its mass use as a disinfectant/bactericide in the consumer market. When BCDMH in reduced slowly in water, dimethylhydantoin is the end product releasing both bromine and chlorine

$$C_5H_6N_2O_2BrCl + 2H_2O \rightarrow HOBr + HOCl + C_5H_8N_2O_2. \qquad (7.1)$$

It is interesting to note that with a moderate reducing agent chlorine is reduced much more slowly than bromine, presumably due to the stronger N-Cl bond of the halogenated amide, so the reaction with a concentrated solution of $Na_2S_2O_5$ is rapid, and leads quantitatively and almost exclusively to the reduction of bromine

$$(7.2)$$

Sulfur dioxide can reduce bromine further to bromide in a method used to prepare hydrobromic acid [3]:

$$SO_2 + Br_2 + 2H_2O \rightarrow 2Br^- + SO_4^{2-} + 4H^+. \qquad (7.3)$$

Hence, hydrogen bromide as well as sulfuric acid (formed in the gas phase) distill over with the water and bromine from the reaction mixture. This results in a loss of product both directly, and due to the fact that the bromide ion greatly enhances the solubility of bromine in the aqueous phase [4]; hence, the amount of metabisulfite used in this experiment is adjusted to be slightly above that required by stoichiometry, at which point a quantitative yield of bromine is obtained. The small amount of HBr that is formed is beneficial in preventing the formation of a bromine clathrate with water of stoichiometry $Br_2.\sim 8H_2O$ [4], which can block the condenser.

7.2   EXPERIMENTAL

A three-neck, 1-L, round-bottom flask equipped with a thermocouple, a 250 mL pressure-equalized dropping funnel, and a still head leading to a 300 mm Graham condenser, is placed on a heating mantle with a temperature controller connected to the thermocouple. The condenser is attached to a receiver feeding a 100 mL round-bottom flask completely immersed in an ice bath. The gas outlet from the receiver is passed over NaOH solution and vented. Due to the low boiling point of bromine (58°C), a single-stage condenser is adequate only if very cold water (~ 4°C) is passed through the coils, and this can conveniently be arranged by immersing a low-power (~2 W) recirculating pump in the receiver ice water bath, which provides the source of cooling for the condenser.

The flask is filled with 350 g of BCDMH powder (16 mesh or finer), and 140 mL of water is added so that the mixture assumes the form of a thick slurry. 104 g of $Na_2S_2O_5$ is dissolved in 200 mL of water and poured into the dropping funnel. The funnel tap is opened and the drip rate adjusted to about 2–3 drops/sec. Red bromine vapor immediately fills the flask, and the bromine, together with a small amount of water starts distilling via the condenser into the receiving flask (Figure 7.1).

With the proper rate of metabisulfite addition, the reaction temperature should reach 83°C–87°C within about 5 min and stabilize, corresponding to conditions under which most of the bromine distills as soon as it is formed. The metabisulfite addition rate must be curtailed if the reaction temperature rises above 90°C, since at this point the metabisulfite decomposes directly into SO_2, which subsequently reduces the distilling bromine to hydrogen bromide. This decomposition is indicated by frothing of the reaction mixture.

When all the metabisulfate has been added and the bromine distillation rate starts to become curtailed, heat is applied to complete the reaction. The temperature is raised to 97°C in about 10 min, and maintained at that value until boiling in the reaction flask ceases and no more bromine distills into the receiver. The vapor in the reaction flask lightens considerably at this stage, and the reaction is complete. A very small amount of bromine can be collected if the temperature is raised further, but it is contaminated with chlorine, and because the vapor is now oxidizing, it tends to form solid clathrate on the condenser coils.

The yield in the receiver is 39 mL of bromine with about 15 mL aqueous phase. This exceeds the stoichiometric yield from Equation 7.2 by a few percent due to ~4% chlorine present in the product in the form of dissolved BrCl. Chlorine can be entirely eliminated by replacing BCDMH with 1-3dibromo-5,5-dimethylhydantoin (DBDMH).

A funnel is used to separate the bromine, sp gr 3.1, from the aqueous layer, which is subsequently shaken thoroughly with an equal volume of 98% H_2SO_4 to remove residual water. This procedure generates some excess pressure, so the funnel need to

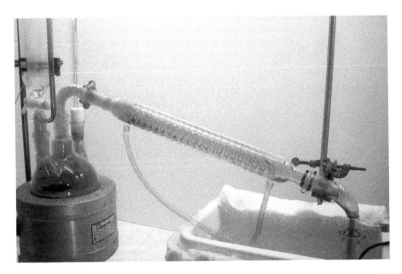

FIGURE 7.1 Sodium sulfite solution at room temperature instantaneously reduces BDCMH generating bromine.

be periodically vented. The bromine forms small oily drops in the acid, which take some time to settle and coalesce. Upon decantation of the lower bromine layer, the top acid layer is found to contain a small amount (~5%) of dissolved bromine.

REFERENCES

1. Clayden, J. P., Greeves, N., Warren, S., and Wothers, P. D., *Organic Chemistry*, p. 442, p. 550. Oxford: Oxford University Press, 2009.
2. Streitwieser, A., and Heathcock, C. H., *Introduction to Organic Chemistry*, pp. 437–8, p. 1187. New York: Macmillan, 1976.
3. Schlesinger, G. G., *Inorganic Laboratory Preparations*, p. 144. New York: Chemical Pub. Co., 1962.
4. Grinbaum, B., and Freiberg, M., Bromine. In *Kirk-Othmer Encyclopedia of Chemical Technology, 5th ed.,* Vol. 4. New York: John Wiley & Sons, 2006.

8 Aluminum Bromide

SUMMARY

- Anhydrous Al_2Br_6 prepared by the union of elements at room temperature.
- Yield of redistilled product is 180 g, or 85% based on Br_2, in a 1 L three-neck flask.

APPLICATIONS

- Difficult Friedel-Crafts alkylations [1].
- Synthesis of alanates from the corresponding hydrides [2].
- Stronger Lewis acid than $AlCl_3$ (decomposes many organic solvents) [1].

8.1 INTRODUCTION

This compound exists as the dimer Al_2Br_6 in the condensed phase. It is a strong Lewis acid, more so than $AlCl_3$, and this together with its dimer molecular structure in both solid and liquid phases, make it more reactive than $AlCl_3$, forming adducts and complexes with most Lewis bases [3]. The increased reactivity makes it a suitable choice when $AlCl_3$-catalyzed reactions are slow or hard to start.

Aluminum bromide is fairly difficult to source commercially, and such material tends to be expensive and often of low reactivity due to extended storage. For these reasons, it is often recommended that the reagent be prepared in situ [4]. There are several preparations available in the literature. Becher [2] describes a setup with a round-bottom, two-neck flask, which is charged with an (unspecified) excess of aluminum turnings placed on a glass wool layer at the bottom. To reduce Br_2 evaporation, bromine is introduced through a capillary reaching almost to the bottom of the flask. The main neck of the flask is connected to an air condenser and carries a side arm for redistillation of the product as well as a gas inlet and outlet through which nitrogen is flushed continuously. Dry Br_2 is run onto the metal at such a rate that Al_2Br_6 refluxes in the middle of the air condenser. The preparation in Schlesinger [5] uses the same reagents, but the reaction vessel is a tubulated retort, and no glass wool layer or refluxing is used; instead, the Br_2 is added at such a rate that the reagents remain liquid (mp Al_2Br_6 98°C, bp 255°C), with heat needed to achieve this toward the end of the reaction.

The present experiment uses a simpler setup and corrects several inaccuracies in both the foregoing preparations. First, due to the high bp of Al_2Br_6, addition of bromine at a rate maintaining Al_2Br_6 reflux in the middle section of the air condenser makes thermal runaway possible. After a period of induction, the reaction of bromine with the aluminum proceeds with explosive violence. Moreover, such reflux leads to excessive loss of bromine through the air condenser and solidification of Al_2Br_6 at the top of the condenser. It is also impossible to use ordinary greased joints. Reflux of bromine is much more desirable; however, this requires an ice water-cooled condenser due to the high temperature differential between the reaction temperature, which must be above the 98°C mp of Al_2Br_6 (190°C was found optimal) and the bp of bromine. Use of such a condenser eliminates the need for introducing the bromine through an extended capillary since bromine is refluxed until entirely reacted. Running the reaction without a reflux condenser is impractical because the bromine reacts with the aluminum in fits and starts, alternating between bright burning flashes and periods of induction because of the protective oxide coating on the aluminum.

The extreme exothermicity of the reaction necessitates the use of a glass wool protective layer at the bottom of the flask, without which the flask may crack at the point where the bromine makes contact with aluminum, due to thermal shock. Continuous flushing with an inert gas is unnecessary, and a single flush at the beginning followed by protection with a drying tube affords good results. Any water vapor present leads to the formation of a solid aluminum oxide/bromide mixture, which is separated during distillation. On the other hand, bromine cannot be separated from the Al_2Br_6 completely in a first distillation (continuous refluxing in the nitrogen stream described by Becher [2] is designed to achieve this). An initial distillation above 200°C at atmospheric pressure passes substantial bromine vapor, as well as some liquid bromine, with the Al_2Br_6 vapor. Subsequent redistillation at ~120°C and aspirator pressure (~20 torr) produces a product that still contains some bromine as evidenced by vapor color and coloration of the solid product. However, after the first ~20% fraction, the remaining condensate is perfectly clear.

8.2   EXPERIMENTAL

Thirty-three grams of aluminum granules is placed in a flat-bottom, 1-L, three-neck flask with a thin, ~1 cm layer of glass wool protecting the bottom. The flask is equipped with a thermometer, a double-surface reflux condenser, and a dropping funnel with pressure equalization containing 185 g bromine, previously dried by shaking with concentrated sulfuric acid. The condenser is cooled by freezing water from a large ice water reservoir with a small (~2 W) recirculating pump. The top of the reflux condenser is protected from moisture by a $CaCl_2$ tube, and prior to commencement of reaction, the setup is flushed with dry nitrogen.

The reaction is started by allowing 10–15 drops of bromine to run into the flask, and waiting up to several minutes for the evolution of white fumes, indicating a reaction. Following this, a few more drops are cautiously added, observing the gradual rise in temperature. The bromine reacts with the aluminum in sporadic fashion, with the addition frequently accompanied by flashes and sparks lasting several seconds; at other times, the addition produces no obvious sign of reaction. With the bromine added in bursts of 5–10 drops, after about 10–15 min the temperature in the flask should reach 210°C–220°C. After about half the bromine has been added, vigorous refluxing commences in the bottom portion of the condenser (Figure 8.1). This provides cooling and allows the addition rate of bromine to be

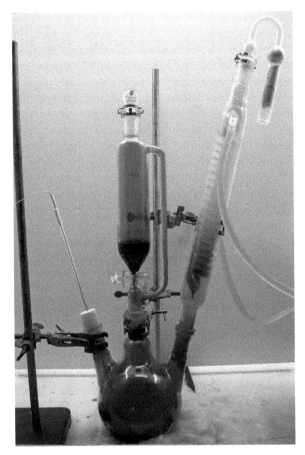

FIGURE 8.1 *A color version of this figure follows page 112.* Bromine reacts with aluminum granules accompanied by sporadic flashing with the temperature rising to 210°C–220°C. The heat volatilizes the unreacted bromine, which refluxes in the condenser until it is all consumed. On cooling, the Al_2Br_6 product crystallizes as white scaly crystals on the flask walls.

increased, controlled by the reflux rate. Provided this regime is maintained, there should be no noticeable escape of bromine at the top of the condenser or deposition of solid Al_2Br_6 in its middle section. The temperature at the end of the reaction drops to just below 180°C; however, no external heating is required. The entire process takes about 2 h.

At the end of the reaction, the insides of the flask are dark red, while white-red flakes are condensed on the flask walls. The reflux condenser is now removed, and the dropping funnel replaced by a downward-sloping distillation head leading to a 250 mL, three-neck receiver flask. One neck of the flask is connected through a small U-tube containing $CaCl_2$ and an empty guard flask to the aspirator vacuum; the other neck is connected to a nitrogen source through an inlet adaptor with a shut-off valve. Vacuum is applied to the system and the reagent flask heated. Initially, bromine vapor fills the apparatus and is removed by the aspirator. This is followed by some liquid bromine that distills when the flask temperature rises above the mp of Al_2Br_6. This is also removed by the aspirator without condensing in the receiving flask. Boiling commences at about 120°C–140°C, and the first fraction collected is of reddish color. After about 10 g of product has passed, the distillate becomes perfectly colorless, and this fraction is gathered. If a multiple feed adaptor is not used and the receiver flask needs to be replaced, the system must be flushed with nitrogen prior to reconnection. Toward the end of the distillation, some Al_2Br_6 "icicles" may be observed sticking to the walls and possibly plugging the adaptors. The glass can be gently warmed at that point with an air gun until the product melts and drops into the receiver flask.

When all product has been collected, the vacuum is disconnected and the apparatus filled with nitrogen. The receiver flask is now disconnected while the product is still liquid and its contents are rapidly emptied into a preheated mortar contained in a nitrogen-filled desiccator, making an effort to spread the liquid onto the mortar walls as much as possible. When the mortar has cooled, the Al_2Br_6 can be ground under nitrogen and the resulting powder stored in a glass-plugged bottle. The yield is about 180 g, or 85% based on the bromine.

Unlike aluminum chloride, aluminum bromide fumes strongly in air even at low humidity. It reacts violently with water, less so with alcohol, which is best used for cleanup, producing copious fumes of white aluminum hydroxide/hydrogen bromide. It is a very strong acid and dehydrating agent and chars most organic compounds, including silicone stop-cock grease after extended contact. THF and diglyme are decomposed almost instantly near room temperature, while the reaction with diethyl ether is very slow below 0°C. It has a tendency to decompose slightly upon storage as evidenced by the appearance of an orange coloration.

REFERENCES

1. Gugelchuk, M., Aluminum bromide. In *Encyclopedia of Reagents for Organic Synthesis*, edited by Paquette, L. New York: John Wiley & Sons, 2004.
2. Becher, H. J., Aluminum. In *Handbook of Preparative Inorganic Chemistry, 2nd ed.*, edited by Brauer, G. p. 806, pp. 813–4. New York–London: Academic Press, 1963.
3. Greenwood, N.N., and Earnshaw, A., *Chemistry of the Elements, 2nd ed.*, pp. 235–6. Oxford: Butterworth-Heinemann, 1997.
4. Fieser, M. and Fieser, L., *Reagents for Organic Synthesis*, Vol. 1, pp. 19–21. New York: John Wiley & Sons, 1969.
5. Schlesinger, G. G., *Inorganic Laboratory Preparations*, p. 14. New York: Chemical Pub. Co., 1962.

9 Lithium Aluminum Hydride

SUMMARY

- Al_2Br_6 and LiH react in Et_2O, yielding 0.68 M $LiAlH_4$ solution at 79% yield.
- Solution loses 27% strength per 24 h with settling of coarse-gray AlH_3 precipitate.
- $LiAlH_4$ formed in situ effects complete conversion of benzonitrile to benzylamine.
- $Ca(AlH_4)_2$ formed by $AlCl_3/CaH_2$ reaction in THF is found to be inert to most organics.

APPLICATIONS

- Powerful, mildly selective, reducing/hydrogenating agent in organic chemistry [1,2].
- Hydrogenates difficult materials, such as B, Si, Zn, and Be [3].

9.1 INTRODUCTION

In addition to its multitude of applications in organic chemistry [1,2], lithium aluminum hydride can also be used to prepare unstable inorganic hydrides, such as B_2H_6, SiH_4, ZnH_2, and BeH_2, an application to which it was originally put by its discoverers [4]. Although fairly readily available, it is quite expensive and often quite impure. Lithium hydride, from which it is easily prepared, is substantially cheaper, much more inert, stores better, and is easier to handle than $LiAlH_4$. Moreover, as demonstrated by the present experiment, the apparatus used for preparing $LiAlH_4$ can often be used without further modification to carry out hydrogenation in situ. Hence, the present laboratory preparation offers some advantages.

$LiAlH_4$ was discovered by Finholt, Bond, and Schlesinger [5] in 1946 following their work on boron hydrides, and culminating in the synthesis of the mixed metal hydrides $LiBH_4$, $NaBH_4$, etc., which possess the strong reducing properties of the boranes while being solid, soluble in many common solvents, and relatively stable.

The aluminum equivalents of the borohydrides turned out to be more difficult to prepare, owing to their lower thermal stability, which made the thermal routes applicable to boron, such as the high temperature reaction of boron halides with hydrogen, unproductive for aluminum. However, a solution of boron trifluoride in ether was found to react with lithium hydride to yield the soluble $LiBH_4$, with the reaction proceeding due to the residual solubility of LiH in that solvent. A similar acid/base reaction between the $AlCl_3$/LiH Lewis acid/base pair was successful in producing an ether-soluble, mixed hydride of lithium and aluminum:

$$4LiH + AlCl_3 \xrightarrow{\text{ether}} 3LiCl + LiAlH_4. \qquad (9.1)$$

The first stage in the reaction is the formation of AlH_3, whose thermal instability makes temperature control essential. However, Reaction 9.1 is quite exothermic due to the high formation energy of LiCl, and it is heterogeneous because of the insolubility of LiCl in ether. Both factors combine to make the process quite difficult to control, as described in the original paper by Schlesinger [4]. Moreover, the original method is quite cumbersome, requiring extended milling of LiH under an inert atmosphere, as well as the use of an "initiator," such as $LiAlH_4$, prepared by more complex means. Other preparations, such as the direct combination of elements [5], are not suitable for the laboratory due to the high pressures (40–200 atm) employed.

Substitution of Al_2Br_6 for Al_2Cl_6 in Reaction 9.1, as described by Wiberg [6,7], eliminates most of the deficiencies of the original method because LiBr is substantially soluble in ether and its enthalpy of formation is lower. Al_2Br_6, being a stronger Lewis acid, also reacts readily with a partially hydrated LiH surface, eliminating the need for an initiator. Thus, a 79% yield is obtained in the experimental section with minimal effort. Because the $LiAlH_4$ product also serves as the initiator and promoter of the reaction, it is possible to conserve Al_2Br_6, by substituting the cheaper Al_2Cl_6, after a small amount of product has formed.

The present experiment produces an ether solution of $LiAlH_4$ saturated with LiBr, with excess LiBr forming a solid precipitate. If it is desired to isolate pure solid $LiAlH_4$, the Al_2Br_6 is dissolved in benzene instead of ether, and the latter evaporated to precipitate LiBr quantitatively [1]. However, this procedure is accompanied by a loss of hydride because a substantial excess of LiH needs to be used to ensure that the AlH_3 monomer formed in the initial stage of Reaction 9.1,

$$3LiH + AlBr_3 \xrightarrow{\text{ether}} 3LiBr + AlH_3.(Et_2O)_n, \qquad (9.2)$$

is completely converted to $LiAlH_4$. Incomplete conversion is of little consequence if the solution is used immediately since the reducing power of the AlH_3 monomer in solution is equivalent to that of $LiAlH_4$; however, unlike $LiAlH_4$, AlH_3 rapidly polymerizes, precipitating granules of the AlH_3 polymer. Thus, in the experimental section, the reducing power of a 0.68 M solution of $LiAlH_4$ and AlH_3 in diethyl ether was found to decrease by 27% over the course of 24 h. Moreover, AlH_3 cannot be separated from ether without decomposition.

In the present experiment, $LiAlH_4$ is not isolated as a pure compound, but used in solution in which it is formed, where the LiBr by-product does not interfere with the reduction. Therefore, as is common practice [8], the yield is calculated on the basis of the solved active hydrogen determined by conversion to hydrogen gas by the action of water on an ether solution:

$$LiAlH_4 + 4H_2O \rightarrow 4H_2 + LiOH + Al(OH)_3. \tag{9.3}$$

The availability of the hydride for reduction is established by its action in reducing benzonitrile to benzylamine [9]:

$$2C_6H_5CN + LiAlH_4 \rightarrow (C_6H_5CH_2N\ )_2AlLi + 4H_2O \rightarrow 2C_6H_5CH_2NH_2 \\ + LiAl(OH)_4, \tag{9.4}$$

where, in agreement with Reference 9, a 1:1 mole ratio results in nearly 100% conversion, with no foreign peaks in the infrared (IR) spectrum (using a smaller mole ratio produces a substantial amount of benzaldehyde) [10].

9.2   DISCUSSION

The solubility of $LiAlH_4$ in ether (29 g per 100 g at 20°C) underpins the covalent nature of its bonding, with the hydrogen atoms forming AlH_4 tetrahedra about a central aluminum atom (bond length 0.155 nm), as well as bridge bonds with lithium atoms (bond length 0.188 nm). $NaAlH_4$ and $KAlH_4$, on the other hand, have a more salt-like structure and hence require more polar solvents, such as THF, to dissolve. In ether solution, $LiAlH_4$ is present in the form of molecular aggregates; at concentrations below about 0.1 M, the dimer predominates, while at 1M the trimer is the dominant species.

Metal aluminum hydrides are stronger reducing agents than the corresponding borohydrides [1]. In particular, they reduce carboxylic acids and their derivatives as well as halides, nitrates, and nitriles. This greater reducing power is partly due to the fact that the hydrogen is more weakly bound in $MAlH_4$, aluminum being more electropositive than boron and hence a weaker electron acceptor, making the hydrogen more readily available to electrophilic attack [1]. Its greater atomic radius (1.48 Å for aluminum as opposed to 0.7 Å for boron) also facilitates the replacement of hydrogen by nucleophilic reagents. Thus, not only do the alanates react more energetically with water and alcohols than borohydrides, they also react with compounds containing much less active hydrogen, such as ammonia, and primary and secondary amines [1].

9.3   EXPERIMENTAL

9.3.1   LITHIUM ALUMINUM HYDRIDE

The ether used in this preparation must be perfectly dry and free of alcohol. This can be achieved by standing with calcium chloride, then with sodium, followed by distillation, the whole procedure taking several days. Alternately, reflux with sodium benzophenone ketyl (10 g sodium and 10 g benzophenone to 1 L of ether) [11] takes only a few hours and also has the advantage of reducing any peroxides formed in the ether, as well as being self-indicating, with a deep blue coloration appearing when the ether is dry. However, if the ether contains more than small traces of water (evidenced by bubbling when clean sodium is first introduced), the sodium can become completely covered by a gray layer of insoluble hydroxide, halting the drying process. This is due to the reduction of water outcompeting the reduction of benzophenone on the sodium surface, with NaOH being insoluble, while sodium benzophenone ketyl is soluble.

To restart the reaction, more sodium can be added or most of the ether temporarily distilled to increase the benzophenone concentration to the point where the sodium surface once again becomes active, as indicated by a coating of blue sodium benzophenone ketyl. The problem can be avoided entirely if the ether is run into the flask containing the sodium and benzophenone in portions, with more ether being added as the contents of the flask become dry. In this way, the ratio of benzophenone to water concentration is maximized. The procedure must, of course, be conducted under inert gas (nitrogen or argon with a bubbler vent).

Aluminum bromide is very soluble in ether, but being a much stronger Lewis acid than aluminum chloride, at high concentrations it reacts with ether at room temperature to form bromides and condensation products in a very exothermic reaction. To avoid decomposition of the solvent, Al_2Br_6 must be added to ether in small amounts at $-10°C$ to $-5°C$ as described by Wiberg [6]. Because it also reacts vigorously with atmospheric moisture, this addition must be performed in a dry atmosphere (flask flushed with inert gas) with subsequent protection provided by a $CaCl_2$ tube, as demonstrated in Figure 9.1.

In a 1 L three-neck flask, equipped with a thermometer, a nitrogen inlet, and a $CaCl_2$ tube, 125 g of dry ether is placed. One neck of the flask is connected through a short 120° bend to a small (~100 mL) flask containing 55 g of coarsely ground Al_2Br_6. The flasks are flushed with nitrogen and placed in an ice–salt mixture at $-16°C$. When the temperature of the ether has dropped to about $-7°C$, the two flasks are tilted so that several grams of Al_2Br_6 drop into the ether. This can be accompanied by effervescence due to localized heating, and the 1 L flask needs to be swirled and rapidly chilled to $-7°C$ to avoid unnecessary decomposition of the ether. This is continued until all the Al_2Br_6 has dissolved, producing a light brown solution. At no stage must the temperature of the ether be allowed to rise substantially above $0°C$ during the addition; otherwise, a dark brown solution substantially contaminated with decomposition products will be obtained. Because the addition is accompanied by evaporation of the ether, some Al_2Br_6 can remain stuck to the walls of the bend of the 100 mL flask. These can be washed down at the end by swirling the ether in the main flask.

Next, 7.6 g of LiH powder, which can contain chunks up to 1–2 mm across, is placed in a flat-bottom flask, located in a water bath equipped with a magnetic stirrer. The hydride is covered with about 42 g of ether, and a dropping funnel with pressure equalizer, containing the previously prepared Al_2Br_6 solution, is connected to the flask. A reflux condenser whose inlet is protected by a $CaCl_2$ tube is now attached. Because the ensuing reflux is rather weak, one can either attach the condenser to the top of the dropping funnel (Figure 9.2) or directly to the flask. The tap of the

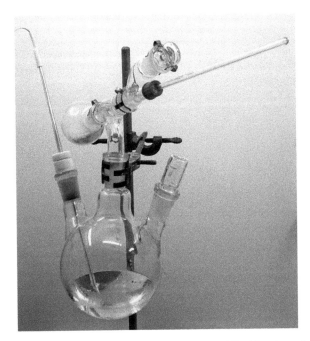

FIGURE 9.1   *A color version of this figure follows page 112.* Aluminum bromide, which decomposes many organic solvents, is here slowly added to diethyl ether with the temperature maintained below 0°C by rapid swirling and cooling the ether in a freezing mixture. The glass stopper is replaced by a $CaCl_2$ protection tube prior to dissolution.

dropping funnel is opened, and the Al_2Br_6 run in at a rate that soon produces gentle reflux of the ether. There is also a small amount of hydrogen effervescence due to decomposition of some of the product. The rate of reflux should remain constant until all the Al_2Br_6 has been added; this takes about a half hour. If excess Al_2Br_6 is added, AlH_3 starts forming by Reaction 9.2, and the reaction becomes much less exothermic. This can serve as a useful indication of an early end point of Reaction 9.1 in cases where the LiH is impure or partly hydrolyzed. The water bath is now heated, and the reactants refluxed for a further 30 min.

All grains of LiH should have disappeared at this stage and the solution should assume a light-gray to dark-gray color depending on the purity of the LiH (dark-colored Li_3N and Al are the main impurities). The solution is allowed to cool and settle, whereupon a layer of very fine LiBr precipitates. If excess Al_2Br_6 was added, some large, gray, irregular crystals of AlH_3 may also begin to form. The maximum yield will be obtained if the $LiAlH_4$ solution is used immediately in situ, which can be done by adding the compound to be hydrogenated (or its solution in ether) through the dropping funnel previously used to add Al_2Br_6. Alternatively, after standing for 15–30 min, the clear liquid can be decanted from the LiBr precipitate. This solution is quite unstable and continues to evolve hydrogen gas, especially when disturbed. It reacts violently with water, and any container into which it is poured must be thoroughly dried to remove surface moisture or else some decomposition will ensue.

FIGURE 9.2 In this in situ preparation of LiAlH$_4$, Al$_2$Br$_6$ solution is slowly added to LiH, maintaining a gentle ether reflux using the moderate exothermicity of this acid/base reaction. Large grains of LiH react and dissolve in about 20 min, forming a LiAlH$_4$ solution, which is extremely sensitive to minute traces of moisture. Decomposition occurs over several days, depositing large grains of polymeric AlH$_3$.

Determination of LiAlH$_4$ content was done by removing 10 mL of the clear solution (unreacted LiH is left behind because it is scantly soluble in ether) and quenching according to Reaction 9.3, by slowly adding water dropwise. This procedure was carried out immediately upon LiAlH$_4$ formation, after 12 h, and after 36 h, yielding 660 mL, 600 mL, and 440 mL of hydrogen gas, respectively. The yield of LiAlH$_4$ based on Al$_2$Br$_6$, immediately upon termination of the reaction is thus 79%. It is evident that the solution loses at least 33% of its reducing power upon standing for 36 h, and this is accompanied by the formation of a 2 cm layer of coarse AlH$_3$ precipitate. Reaction with benzonitrile in the next section shows that the decrease in reducing power of the original LiAlH$_4$ solution is even greater than indicated by quenching with water, which is likely due to an extensive amount of unreactive AlH$_3$ polymer in solution, which has not yet precipitated.

9.3.2 REDUCTION OF BENZONITRILE WITH LiAlH$_4$

To establish the reducing power of the LiAlH$_4$, ether solution was used to reduce benzonitrile to benzylamine, in accordance with a literature procedure [9] reporting a high yield (85%) with a 1:1 mole ratio of the reagents. Accordingly, 5.85 g (0.057 mol) of benzonitrile was run in over the course of 15 min onto 100 mL of the LiAlH$_4$ solution (0.0675 mol). The self-refluxing mixture was protected by nitrogen and magnetically stirred until all benzonitrile has been added. At this stage, reflux was continued by heating on a water bath for a further 30 min.

The procedure for hydrolyzing the benzylamine–LiAl complex so formed, as given in the literature [9], was found to lock the product into a stiff gel from which it was difficult to extract; hence, it was modified as follows. The excess $LiAlH_4$ was decomposed with 2 mL of water, followed by 10 g NaOH dissolved in 40 mL of water. The flask contents were shaken and more NaOH solution added in small increments as required, until all inclusions of a gray-green gel containing the product were broken down. At this stage, a light-yellow ether layer should separate from the insoluble salts and is decanted. The salts are further washed with a small quantity of ether, and the portions combined. The ether is removed in a water bath, leaving a clear-slightly yellow solution. This can be dried in an air oven at 120°C or with Na_2SO_4. A thin film IR spectrum showed the product to be benzylamine, with no foreign peaks at the 2% level.

9.3.3   CALCIUM ALUMINUM HYDRIDE

This compound was prepared using the apparatus in Figure 9.2, in accordance with the original article [12] (with a translation given by Becher [7]), where a 60% yield was reported. In contrast to some reports [1], the author has found that CaH_2 reacts with Al_2Cl_6 in tetrahydrofuran readily and extensively, as evidenced by the substantial heat evolution accompanying the process. However, the degree to which the resulting product $Ca(AlH_4)_2$ is capable of reducing organics has apparently not been investigated in the literature.

Calcium aluminum hydride is reported to be soluble in THF (0.3 mol/L [1]); however, here there was no observable solubility, with the solvent inactive to water, while most of the alanate forms a solid complex with THF, which gradually settles to the bottom. The precipitate, which does react vigorously with water, contains up to 60% THF from which $Ca(AlH_4)_2$ can only partially be separated even under aspirator vacuum at 90°C [7]. No evidence has been found that either the solution or the complex is capable of reducing organic compounds despite numerous attempts with various aldehydes and nitriles, which $LiAlH_4$ reduces readily and completely. It appears this is due to the insolubility of calcium aluminum hydride and the very strong complexing action of THF, possibly involving partial reduction of the latter as mentioned by Schlesinger and Wartik [13]. In any case, the THF cannot be fully separated from this complex without decomposition. It seems the stability of the complex makes its participation in reduction energetically unfavorable.

REFERENCES

1. Hajos, A., *Komplexe Hydride und ihre Anwendung in der Organischen Chemie*. Berlin: Deutscher Verlag der Wissenschaften, 1966.
2. Hochstein, F. A., Quantitative studies on lithium aluminum hydride reactions. *J. Am. Chem. Soc.* 71(1): 305–7, 1949.
3. Greenwood, N. N. and Earnshaw, A., *Chemistry of the Elements, 2nd ed.*, pp. 229–30. Oxford, U.K.: Butterworth-Heinemann, 1997.

4. Finholt, A. E., Bond, A. C., and Schlesinger, H. I., Lithium aluminum hydride, aluminum hydride and lithium gallium hydride, and some of their applications in organic and inorganic chemistry. *J. Am. Chem. Soc.* 69(5): 1199–1203, 1947.
5. Clasen, H., Alanat-Synthese aus den Elementen und ihre Bedeutung. *Angew. Chem.* 73(10): 322–331, 1961.
6. Wiberg, E. and Schmidt, M., Über eine Vereinfachte Darstellung des Lithium-Aluminium-Wasserstoffs LiAlH$_4$, *Zeit. Naturforsch.* 7b: 59–60, 1952.
7. Becher, H. J., Aluminum. In *Handbook of Preparative Inorganic Chemistry, 2nd ed.,* edited by Brauer, G. pp. 806–7. New York–London: Academic Press, 1963.
8. Johnson, J. E., Blizzard, R. H., and Carhart, H. W., Hydrogenolysis of alkyl halides by lithium aluminum hydride. *J. Am. Chem. Soc.* 70(11): 3664–5, 1948.
9. Amundsen, L. H. and Nelson, L. S., Reduction of nitriles to primary amines with lithium aluminum hydride. *J. Am. Chem. Soc.* 73(1): 242–244, 1951.
10. Soffer, L. M. and Katz, M., Direct and reverse addition reactions of nitriles with lithium aluminum hydride in ether and in tetrahydrofuran. *J. Am. Chem Soc.* 78(8): 1705–9, 1956.
11. Furniss, B. S., Hannaford, A. J., Smith, P. W. G., and Tatchell, A. R., *Vogel's Textbook of Practical Organic Chemistry, 5th ed.,* pp. 405–6. London: Addison Wesley Longman, 1989.
12. Schwab, W. and Wintersberger, K., *Zeit. Naturforsch.* 8b: 690, 1953.
13. Schlesinger, H. I. and Wartik, T., Hydrides and Borohydrides of Light Weight Elements and Related Compounds. Armed Services Technical Information Agency Report, AD97410. http://www.dtic.mil/cgi-bin/GetTRDoc?AD=AD097410&Location=U2&doc=GetTRDoc.pdf

10 Triethylaluminum and Diethylaluminum Bromide

SUMMARY

- Grignard-type reaction of magnalium/EtBr gives $AlEt_2Br$ in 94% yield.
- $AlEt_2Br$ reduction with sodium gives 72% yield of $AlEt_3$, containing ~4% $AlEt_2Br$.
- A simple EtBr preparation suitable for the present synthesis is described.
- A laboratory preparation of magnalium from Al and Mg is presented.

APPLICATIONS

- Ziegler–Natta alkene polymerization catalysts [1,2].
- Alkylation promoters for weak Lewis bases [3].
- Precursors to organometallics with less electropositive center (e.g., Sn, Pb, Ga, B [4]).
- Effective H_2O and O_2 scavengers.
- Initiators of hydrogenation reactions (e.g., $NaAlH_4$ [5,6]).

10.1 INTRODUCTION

These pyrophoric organoaluminum liquids are very efficient catalysts in a wide variety of polymerization reactions and also have found use as hydrogenation initiators (e.g., formation of $NaAlH_4$ or $AlEt_2H$ under pressure from the elements [5,6]). Because these compounds ignite spontaneously on contact with oxygen or moisture, there are restrictions on their transportation. Here, a simple method for their preparation in the laboratory is presented based on several procedures reported in the literature, which have been adapted to save time, simplify apparatus, and in certain cases improve the yield.

Compounds of the form AlR_nX_{3-n}, where X is a halogen and R an alkyl or alkoxy group, were first prepared and analyzed by Grosse and Mavity in 1939 [7], but their importance in chemistry was not established until the pioneering work of Ziegler and Natta [1], which demonstrated their use as very effective alkene polymerization catalysts. There are two types of polymerization processes: the polymerization of ethene into unbranched chains up to C_{200} by consecutive addition to $AlEt_3$ at ~100 atm, and the low-pressure polymerization of unbranched terminal alkenes by treatment with a mixture of $TiCl_4$-$AlEt_3$ or $TiCl_3$-$AlEt_2Br$ (Ziegler–Natta catalyst) in a low-bp hydrocarbon. Such a mixture is reactive enough to absorb and polymerize ethene even under standard conditions.

10.2   DISCUSSION

The industrial method for the preparation of aluminum trialkyls [1] relies on the action of hydrogen on aluminum and AlR_3 to produce the monohydride AlR_2H, followed by addition of C_2H_4 to form more AlR_3 at pressures of about 200 atm.

More suitable for the laboratory are the methods for producing AlR_nX_{3-n} based on the oxidation of aluminum by an alkyl halide [7],

$$3RX + 2Al \rightarrow R_2AlX + RAlX_2, \tag{10.1}$$

which results in an equimolar mixture of the mono- and di-halides. This reaction, being similar to the Grignard reaction, requires extremely dry apparatus and solvents in order to commence. In addition, the products react spontaneously with oxygen, and thus the preparation must be conducted under nitrogen or argon. Iodine, or a small amount of previously prepared product, is added as initiator and works by attacking the protective oxide layer of aluminum and magnesium and forming reactive intermediates. Even then, an induction period of several days is sometimes required [7]. The products are generally not distributed strictly according to Reaction 10.1, but form a mixture of variable composition, and disproportionation is possible:

$$2RAlX_2 \rightarrow R_2AlX + AlX_3. \tag{10.2}$$

Thus, C_2H_5Cl acts on aluminum turnings in an autoclave at room temperature yielding 59% ethylaluminum dichloride. Fractionation is carried out at aspirator pressure (50–100 torr) in dry nitrogen and requires an efficient Podbielniak column.

These reactions have a tendency to thermal runaway due to a vigorous decomposition to saturated alkanes and tarry condensation products [7]. As seen in the $LiAlH_4$ preparation, the use of bromides is preferable in this respect to chlorides because of the lower heat of formation of the product alkali halide, making the reaction easier to control. With bromides, separation by a single Podbielniak distillation is ineffective, but according to Grosse and Movety, the dialkylaluminum bromide can be prepared directly by reducing the dibromide component in Reaction 10.2 with a stoichiometric amount of magnesium alloyed to aluminum,

$$(2Al + Mg) + 4RBr \rightarrow 2R_2AlBr + MgBr_2, \tag{10.3}$$

which corresponds closely to the commercially available magnalium alloy (70% Al, 30% Mg). This reaction carried out in a 1-L, three-neck flask with mechanical stirring at 120°C–140°C [7] takes 3 h and yields 91% of a 98% pure product (the main impurity being the dibromide).

In an alternative preparation [8], the reagents in 10.1 are refluxed at 200°C–220°C for several hours with a slight excess of anhydrous sodium bromide, which complexes the Al_2Br_6 formed in the spontaneous disproportionation of $AlRBr_2$ according to

$$2RAlBr_2 + NaBr \rightarrow R_2AlBr + AlBr_3.NaBr. \tag{10.4}$$

Distilling at 3 mm and 84°C gives a 91% yield of essentially pure $AlEt_2Br$.

Triethylaluminum bromide is prepared from diethylaluminum bromide by reduction with a slight excess of sodium (a large excess must be avoided as the reduction can proceed further to the quaternary compound):

$$3AlEt_2Br + 3Na \rightarrow 2AlEt_3 + Al + 3NaBr. \qquad (10.5)$$

Grosse and Mavity describe this as a 17-h reaction to a final temperature of 200°C–210°C, giving a 62% yield of $AlEt_3$ contaminated with 11% of $AlEt_2Br$. Repetition of the procedure with a stoichiometric amount of sodium calculated to complete the reaction gives an overall 60% yield of 99% $AlEt_3$ based on hydrolysis.

Here, it is shown that the literature results [7,8] can be easily reproduced with simpler apparatus and shorter reaction times. In particular, specialized Schlenklines can be eliminated without introducing any substantial problems or penalties in yield. Long induction periods in the reaction of aluminum with EtBr can be avoided by employing a high-temperature reflux, which quickly starts the reaction even with dull metal turnings and a significant alcohol impurity in the reagent. Thermal runaway is entirely absent if excess ethyl bromide is kept to a minimum by continuously refluxing the reagent into a reservoir. This eliminates the possibility of flooding the condenser and keeps a fairly constant and easily controlled temperature in the reaction mixture. Refluxing excess reagent containing traces of the alkyl aluminates also has the beneficial effect of drying the contents of the reservoir prior to their introduction into the reaction flask. A 94% yield of $AlEt_2Br$ based on EtBr is obtained here, with the product essentially pure based on hydrolysis.

Reduction with sodium in Reaction 10.5 was found to proceed to completion on a time scale of several minutes after the sodium melted, rather than hours. The reaction is very exothermic, resulting in vigorous boiling (T >250°C), which can lead to decomposition of the organometallic. For this reason, cooling must be rapidly applied as soon as excessive reflux commences and an air condenser sufficient to accommodate the boiling alkyl aluminum is used. Since the reaction is two-phase, the sodium globules become covered by a porous coat of fine Al/NaBr of large surface area and the consistency of pumice, which fills the reaction flask and makes the mixture unstirrable. Small amounts of sodium are trapped inside such aggregates, which prevent further reaction. Because the amount of sodium used must be stoichiometric to avoid further reduction, the exact amount required is difficult to establish, necessitating a repeat reduction [7,9]. The amount of unreacted sodium (which attacks the glass upon distillation) can be minimized by vigorous stirring immediately prior to the mp of sodium, which serves to disperse it as much as possible and avoids formation of large Al/NaBr aggregates. The reaction and distillation can be carried out in the same apparatus by closing off the nitrogen source and replacing the bubbler by a vacuum line. The first-stage reduction in the present experimental procedure gives a 76% yield of $AlEt_3$ with a 15% $AlEt_2Br$ impurity based on hydrolysis. Repeat reduction with a slight excess on sodium is much less exothermic and produces $AlEt_3$ at 72% yield for both steps, containing 4% $AlEt_2Br$.

Since this preparation is based on the fairly uncommon 30% Mg–70% Al alloy, an experiment was conducted to see whether a mixture of the same composition can effect the conversion in Reaction 10.3. It was found that in this case the reaction proceeds completely differently, with the two metals reacting essentially independently with the ethyl bromide, so that the aluminum alkyls are distributed essentially according to Reaction 10.1. For this reason, a simple, efficient method for preparing a Mg–Al alloy in suitable form is also presented here.

Ethyl bromide is readily available, but it is not often stocked in the laboratory and can be prepared in under an hour. Because the few preparations available in the literature generally contain extensive purification stages [10] that the author has found are unnecessary here, the experimental section commences with the preparation of this reagent for the present experiment.

10.3   EXPERIMENTAL

10.3.1   ETHYL BROMIDE

The following proportions give maximum yield of EtBr (bp 38°C) in a 1-L flask without foam spilling into the receiver. In a 1-L, round-bottom, single-neck flask, 500 g of 98% H_2SO_4 is placed, and 225 g of 95% ethanol is added in portions with swirling, but without cooling, to aid the formation of ethyl hydrogen sulfate. After cooling the mixture in an ice bath, 185 mL of ice-cold water is added with swirling, followed by 253 g of powdered NaBr, and a few boiling chips. Next, the flask is fitted with a bent antisplash head connected to a 300 mm Graham condenser leading into a 250-mL receiver through a distillation adapter with vented take-off. The Graham condenser coil is cooled with ice water circulated by means of a small (2 W) recirculating pump from a reservoir in which the receiver flask is immersed.

The reagent flask is heated on a 300 W mantle, whereupon the NaBr, which initially forms a hard cake at the bottom of the flask, rapidly dissolves and soon a steady stream of an oily liquid and a small amount of water starts to distill, with the distillation head temperature reaching about 45°C. In the initial stages of the distillation, foam rapidly builds up, and the heating rate must be carefully regulated to avoid flooding the splash head. Ideally, the foam height should be kept below 2 cm during the distillation. As the foam subsides, the heating power is gradually increased until the flask temperature reaches 92°C, at which point the distillate coming over is almost entirely water, and its amount is greatly diminished. The clear mixture in the receiver is separated in a 250-mL funnel, yielding about 250 mL or 297 g in the bottom $EtBr/Et_2O/EtOH$ layer, and about 50 mL in the top $H_2O/HBr/EtOH$ layer.

Purification for present purposes consists of removing the water and most of the ethanol from the bottom layer with concentrated H2SO4. The 98% acid decomposes the EtBr somewhat with release of bromine indicated by the appearance of a yellow coloration and the formation of solid condensation products. Therefore, slightly weaker 83% acid is used instead. Two hundred grams of 98% H2SO4 is diluted with 40 mL water, cooled in an ice bath and shaken with the EtBr fraction. During shaking, the flask is periodically cooled in an ice bath, and pressure carefully released. Upon separation the top organic layer is reduced to 236 g. This is placed in a 250-mL, single-neck flask and stirred for 30 min with 10 g of fused and crushed $CaCl_2$. Then, some boiling chips are added, the flask is placed in a water bath, and the EtBr distilled in a thoroughly dried apparatus, with the bath temperature reaching 47°C at the end. The yield is 230 g EtBr (86% based on NaBr), sp gr 1.46, and the product still contains a few percent EtOH, which has no effect on the following reactions.

10.3.2 Magnalium

Magnesium autoignites in air just above its melting point; hence, it is essential that the protective fused salt layer described below is not dispensed with. A stainless steel tube, 27 mm inside diameter and 130 mm long, closed at one end, is filled with 85 g of a thorough mixture of 30% Mg turnings 70% Al granules. The mixture is pressed into the tube so as to leave 3 cm free at the end. Next, a layer about 1 cm deep of a 50:50 $NaCl:CaCl_2$ mixture is added, and the tube is filled with fine washed sand up to about 1 cm from the top. The tube is now clamped to an approximately 60 cm long S/S rod attached to a vibrating device and lowered into a top-loading oven.

The oven temperature is raised to 640°C in the course of about 30 min, at which point the vibrating device is started, and then it is raised more slowly to 750°C in about 45 min. The salt mixture melts at about 600°C and, combined with the sand, protects the magnesium, mp 649°C, and aluminum, mp 650°C, from oxidation. If any sparking is evident from the mouth of the tube, more sand needs to be added.

When the tube has cooled, a rotating wire brush placed inside the tube mouth rapidly removes the salt/sand cover revealing a polished metal surface. The tube is now clamped in a drill press and a 25.4-mm (1-in.) drill rotating at a low speed is used to drill out the metal, producing a fine gray/black magnalium powder.

10.3.3 Diethylaluminum Bromide

A single-neck, flat-bottom, 250-mL flask containing 40 g of magnalium powder, about 0.1 g iodine, and a 3-cm stir bar is placed in a paraffin bath equipped with a heater/stirrer. A 250-mL dropping funnel with pressure equalization and charged with 180 mL EtBr is attached to the flask, and a 300-mm, double-surface, reflux condenser that is coil cooled by recirculating ice water is connected on top. The outlet of the reflux condenser is connected to an inert gas source and bubbler vent.

The apparatus is flushed with inert gas for several minutes, and then EtBr is run onto the magnalium powder and iodine until they are just saturated. A reaction does not usually start at this point; hence, the bath heater is turned on and the EtBr evaporated back into the dropping funnel with a bath temperature of about 65°C being necessary at the end. At this stage, white fumes of $Al(OH)_3$ appear due to the reaction of alkylaluminate with residual water/alcohol, and this indicates the start of reaction. If this does not occur, the bath temperature can be raised a further 10°C–20°C and EtBr run in at the same rate at which it is being refluxed back into the dropping funnel (Figure 10.1). This starts the reaction very quickly.

The alkylaluminate fumes carried over into the dropping funnel rid the reagent of the last traces of water and alcohol, and also react with the iodine initially carried over into the funnel containing the EtBr, turning the liquid colorless as the iodine is

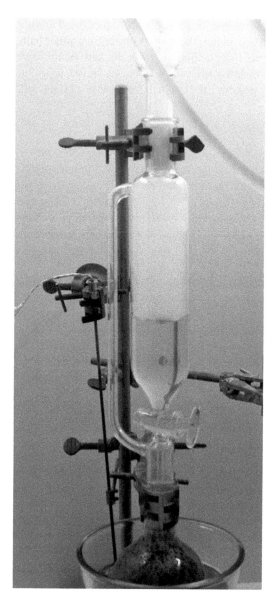

FIGURE 10.1   Ethyl bromide is run onto magnalium powder and iodine at about 65°C. Some alkylaluminum generated refluxes back into the dropping funnel together with unreacted ethyl bromide. This dries the reagent of residual water/alcohol and produces white fumes of $Al(OH)_3$ indicating the start of reaction.

reduced. The EtBr is now dripped in at a rate of about 2–3 drops, while the bath temperature is maintained at 65°C, and the unreacted EtBr is being recycled back into the dropping funnel. The amount of EtBr in the funnel gradually decreases indicating the progress of the reaction, until after about 90 min all has reacted and no more reflux occurs. The contents of the flask are gel-like at this stage so that magnetic stirring usually ceases. The bath temperature is now raised to about 110°C in 20 min to complete the reaction.

The flask is now removed from the bath and cooled to about 0°C–10°C, at which point the vapor pressure of the alkylaluminates is sufficiently low that the flask can be disconnected from the apparatus and exposed to the atmosphere without risk of the contents catching fire. Upon disassembly and exposure to the atmosphere, traces of ethylaluminates clinging to the apparatus walls fume strongly, producing a white coat of aluminum hydroxide, but they do not ignite, due to the small volume of the reagent. However, spills of more than a few tenths of a milliliter ignite instantly.

The flask is now placed on a heating mantle, and dry apparatus arranged for downward distillation in the vacuum of a single-stage vacuum pump (this is nominally 0.1 torr at the pump; however, due to the flow resistance of the apparatus, the actual pressure is about 2–3 torr). The condenser coil and receiver are cooled in ice water since the bp of the lowest boiling ethylaluminate, $AlEt_3$, is 25°C at 1 torr. A dry ice or liquid nitrogen trap is normally employed ahead of the pump; however, these are not strictly necessary as the vapor pressure of residual ethylaluminates in the vacuum line with the above arrangement is so low that a connection to the vacuum pump through several meters of hose can suffice without any ill effects to the pump or its residual pressure.

Liquid starts to come over at about 70°C, and heating needs to be gentle in the beginning to prevent violent bumping, as the product is dispersed in a gel. At the end, the temperature is raised to 110°C to fully dry the flask, with 127.4 g, sp gr 1.35, being collected in the receiver. Hydrolysis of 1.65 g of the product released 495 mL of insoluble gas at 30°C, compared to the expected 493 mL for ethane from $AlEt_2Br$, showing the product is essentially pure. The yield is thus 94%.

10.3.4   TRIETHYLALUMINUM

116.4 g of $AlEt_2Br$ in a 250-mL, round-bottom flask is cooled in an ice bath to reduce the vapor pressure of the alkylaluminum, and then 18 g of sodium in 1 g pieces is added quickly, after being washed in ether to remove the paraffin. A 3-cm stir bar is then inserted and a 300-mm air reflux condenser attached to the flask to accommodate any frothing. The top of the reflux condenser is connected through a 180° bend to a Graham condenser arranged for downward distillation into a 250-mL receiver cooled in an ice bath. An adapter carrying a gas inlet and bubbler outlet is inserted between the condenser and the receiver flask. The whole apparatus is then flushed with inert gas before the flask containing the reagents is quickly attached to the air condenser, minimizing ingress of air, which manifests itself as a white mist of aluminum oxide/hydroxide, which fills the flask as soon as stirring is commenced.

The reagent flask is placed in a paraffin bath that is quickly heated with brisk stirring. At the melting point of sodium, a reaction manifests itself by the appearance of small black particles on the walls of the flask. At about 107°C bath temperature, the reaction becomes extremely vigorous with the mixture boiling over into the air condenser and the reagent temperature rising above 150°C. Chilled paraffin is quickly added to the bath briefly decreasing the bath temperature below 90°C and the boiling becomes more measured. Within about 5 min, the flask fills with large black porous pieces consisting of an aluminum–sodium bromide mixture, each evidently originating from individual sodium pieces (Figure 10.2). The flask temperature is now raised to 150°C about 20 min. There are no signs of reaction at this stage; however, the level of liquid in the flask diminishes somewhat as it is soaked up into the Al–NaBr aggregate. The heating at the end is designed to consume some of the unreacted sodium that might be present inside the Al–NaBr matrix.

The paraffin bath with flask is now cooled to below 30°C, at which point the inert gas inlet is shut off and the bubbler replaced by a vacuum line from an oil pump. Vacuum is applied, with the pressure dropping to 300 mtorr at the pump and the paraffin bath heated with liquid starting to come over at 45°C. The temperature has to be raised to 90°C at the end, which dries the black lumps, resulting in their color changing to a dull gray with a white crust of NaBr. The temperature cannot be raised further as this results in the lumps cracking violently, releasing small particles

FIGURE 10.2 *A color version of this figure follows page 112.* The reaction between sodium and Et_2BrAl commences at the melting point of sodium and becomes extremely vigorous at about 107°C. After a few minutes, the flask fills with large, black, porous pieces of Al–NaBr aggregate that soak up the Et_3Al product.

of sodium, which attack the glass. The vacuum is shut down, and nitrogen let in to atmospheric pressure.

The yield is 48.1 g of a clear viscous liquid, of which 0.93 g was hydrolyzed to give 540 mL of insoluble gas. Assuming that no quaternary aluminum salt is present (this is nonvolatile), the product corresponds to 87% triethylaluminum, whose yield is therefore 76%, with 13% diethylaluminum bromide impurity. The results agree with the literature [7,9], but with a higher yield.

At slightly above the stoichiometric requirement of 1.2 g, 1.5 g of ether-washed sodium in 0.2 g pieces was now added to 46 g of the above mixture and heated as before, but with very rapid stirring commencing at the melting point of sodium. This time the sodium was well dispersed prior to reaction so the deposited aluminum and NaBr assumed the form of a fine black powder, and the mixture remained stirrable throughout. After 20 min stirring with a maximum temperature of 115°C, the reagent flask was again cooled below 40°C, and the resulting distillation produced 39 g of liquid, which came over at 45°C–50°C. Hydrolysis of 0.43 g of the liquid released 290 mL of gas corresponding to essentially pure $AlEt_3$ (278 mL calculated for ethane at 30°C). The combined yield is therefore 72%. An IC analysis, Appendix, Figure A.5, shows the presence of ~4% $AlEt_2Br$.

10.3.5 DIETHYLALUMINUM BROMIDE FROM ETHYLALUMINUM SESQUIBROMIDE

Ethylaluminum sesquibromide traditionally denotes an equimolar mixture of $AlEtBr_2$ and $AlEt_2Br$, resulting from the reaction of aluminum with ethyl halides in Reaction 10.1. In the present case, it was produced in an experiment designed to establish whether magnalium can be replaced in Reaction 10.3 by a mixture of the elements. This reaction differed from that of the alloy in that the reagents stayed fluid and stirrable throughout; 7.7 g Al and 3.3 g Mg reacted with 47.3 g EtBr to give 45.9 g of a viscous liquid, 1.60 g of which hydrolyzed to yield 310 mL of insoluble gas at 30°C, with 309 mL calculated for an equimolar mixture. The yield is thus 85%, with a substantial portion of residual EtBr presumably lost in forming a Grignard reagent with the magnesium

Because the synthesis of diethylaluminum described above uses a magnesium–aluminum alloy with 30% Mg content that often needs to be prepared, it is of interest to establish whether this can be replaced by a two-stage procedure relying on the disproportionation of ethylaluminum dibromide and complexing of the aluminum bromide so formed by NaBr, as described by Zakharkin and Khorlina [8].

With stirring continuing for 2 h, 44.3 g ethylaluminum sesquibromide was heated with 18 g of NaBr (slight excess) at 190°C–200°C. On cooling, a white crystalline mass separated below a clear liquid. The flask contents were now distilled in an oil bath in the vacuum of a single stage oil pump (0.3 torr at the pump, and about 2 torr in the reactor flask), with the volatile product coming over at 50°C–52°C. The oil bath temperature was briefly raised above 95°C to melt the lower $AlBr_3.NaBr$ layer and release the last drops of product; no distillate came over when the temperature was raised a further 20°C. Upon cooling, the contents of the reactor flask solidified to a gel, and 18.8 g of liquid product gathered in the ice-cooled receiver. This represents a yield of 66% based on the sesquibromide, which is below the 91% yield reported by

Zakharkin and Khorlina [8]. The 0.91 g of product was hydrolyzed releasing 268 mL of insoluble gas (267 mL theoretical for diethylaluminum bromide). Several grams of product were lost clinging to the walls of the distillation apparatus with the remainder left in the gel in the reaction flask.

It is possible to conduct simple operations with alkylaluminum halides without resorting to Schlenk-type equipment, and the extra time this requires. Thus, these reagents can be pipetted between flasks if the source flask is preliminarily cooled in an ice-salt bath to minimize the reagent vapor pressure, and the receiver flask is preliminarily heated to drive out moisture clinging to the walls and flushed with nitrogen to remove oxygen. Provided the greased stoppers to the flasks are opened only long enough to fill and empty the pipette, and the pipette is transferred rapidly between flasks, very little decomposition occurs, as evidenced by the amount of fumes produced. Should the reagent be introduced into an unprepared flask, much heating, fuming, and charring occur.

REFERENCES

1. Ziegler, K., Gellert, H.-G., Zosel, K., Holzkamp, E., Schneider, J., Söll, M., Kroll, W.-R., Metallorganische Verbindungen, XXXIV Reaktionen der Aluminium-Kohlenstoff-Bindung mit Olefinen. *Eur. J. Org. Chem. (Ann. Chem.)* 629(1): 121–166, 1960.
2. Kim, J. S., Wojcinski, L. M., Liu, S., Sworen, J. C., and Sen, A., Novel aluminum-based, transition metal-free, catalytic systems for homo- and copolymerization of alkenes. *J. Am. Chem Soc.* 122(23): 5668–9, 2000.
3. Suzuki, K., Nagasawa, T., and Saito, S., Triethylaluminum. In *Encyclopedia of Reagents for Organic Synthesis*, edited by Paquette, L. New York: John Wiley & Sons, 2004.
4. Greenwood, N. N. and Earnshaw, A., *Chemistry of the Elements, 2nd ed.*, p. 260. Oxford, U.K.: Butterworth-Heinemann, 1997.
5. Ashby, E. C., A direct route to complex metal hydrides. *Chem. Ind.* 208–9, 1962.
6. Clasen, H., Alanat-Synthese aus den Elementen und ihre Bedeutung. *Angew. Chem.* 73(10): 322–331, 1961.
7. Grosse, A. V. and Mavity, J. M., Organoaluminum compounds. *J. Org. Chem.* 5(2): 106–121, 1940.
8. Zakharkin, L. I. and Khorlina, I. M., Improved method for the preparation of diethylaluminum hydride. *Russ. Chem. Bull.* 9(1): 133–134, 1960.
9. Becher, H. J., Aluminum. In *Handbook of Preparative Inorganic Chemistry, 2nd ed.*, edited by Brauer, G. p. 810. New York–London: Academic Press, 1963.
10. Gattermann, L., *Laboratory Methods of Organic Chemistry*, pp. 95–6. Revised by Wieland, H. New York: The Macmillan Co., 1937.

11 Hydrazine Sulfate and Alcoholic Hydrazine Hydrate

SUMMARY

- Hofmann degradation of urea yields 62% $N_2H_6SO_4$.
- IC analysis gives 95% $N_2H_6SO_4$, 3% NaCl, and 2% Na_2SO_4 on first workup.
- Metal ions are complexed with EDTA to improve product purity.
- Yield loss accounted for by N_2 evolution with 23% occurring in the oxidation stage.
- Hydrazine as $N_2H_4.2H_2O$ obtained by low-temperature alcoholic extraction in 75% yield.

APPLICATIONS [1]

- Synthesis of hydrazones and azines from carbonyl compounds.
- Wolff–Kishner decomposition of hydrazones to alkanes, and dihydrazones to alkynes.
- Heterocycle synthesis, for example, Einhorn–Brunner formation of triazoles [2].
- Gabriel amine synthesis with hydrazine-assisted cleavage of phthalimide intermediate.
- Diazotization of the N-N moiety to yield azides [3].
- Reduction of nitro compounds to amines.
- Phenylhydrazine derivatives used in characterization of carbonyl compounds.
- Convenient reducing agent/antioxidant, H_2O, and N_2 are the only products [4].

11.1 INTRODUCTION

The fact that hydrazine hydrate can be prepared in high yield using cheap materials in a few hours in the laboratory is not generally well known and, hence, its preparation is presented here. Hydrazine is often introduced into a reaction in the form of its sulfate if an anhydrous reaction medium is required [5], or due to the much reduced air sensitivity of the sulfate. The synthesis of this salt, therefore, is also presented.

11.2 DISCUSSION

11.2.1 HYDRAZINE SULFATE

The first hydrazine-class compound synthesized was phenylhydrazine by Emil Fischer in 1875, by diazotizing aniline and reducing the resulting diazonium salt with sulfite. Hydrazine hydrate was first prepared by Curtius in 1887, by condensing ethyl diazoacetate in concentrated alkaline solution, followed by acidic hydrolysis. The first commercially successful process for hydrazine production, still used today, was introduced by Raschig in 1907 [6] and consists of a low-temperature (5°C–10°C) oxidation of ammonia by sodium hypochlorite solution to chloramine,

$$OCl^- + NH_3 \rightarrow NH_2Cl + OH^-, \tag{11.1}$$

followed by a rapid high-temperature (~130°C) and high-pressure (~100 atm) reaction with anhydrous ammonia to produce hydrazine:

$$NH_2Cl + NH_3 + OH^- \rightarrow N_2H_4 + H_2O + Cl^-. \tag{11.2}$$

The extreme conditions of the chloramine/ammonia reaction are necessary because chloramine oxidizes ammonia much more slowly (~20-fold) than hydrazine (itself being reduced to ammonia) in a reaction that is extremely sensitive to the presence of catalysts,

$$NH_2Cl + N_2H_4 \rightarrow N_2 + 2NH_4^+ + 2Cl^-, \tag{11.3}$$

so that very low concentrations (~10^{-5} M) of transition metal ions such as Cu^{2+}, increase the speed of the reaction by several orders of magnitude [6]. For this reason, a large excess (~20-fold) of ammonia is continuously recycled, and glue or gelatine is used to bind impurity ions in a colloidal suspension.

Several literature preparations [7,8] describe a laboratory-scale Raschig process for the preparation of hydrazine sulfate using gelatine as ion binder [7], with a yield of 20% based on hypochlorite, and correspondingly only 1% based on ammonia, since the latter is not recycled. Even smaller yields are obtained using Mn^{2+} ions as catalyst inhibitors [9].

The present laboratory method for preparing hydrazine sulfate is based on the Schestakow reaction [10], which is a Hofmann-type degradation of urea by the action of a hypochlorite salt in basic conditions, with the acyl group converted to carbonate:

$$NH_2CONH_2 + OCl^- + 2OH^- \rightarrow N_2H_4 + CO_3^{2-} + H_2O + Cl^-. \tag{11.4}$$

Although this reaction is formally analogous to Hofmann degradation of primary amides $RCONH_2$, with R replaced by the amine moiety NH_2, it is uncertain whether it proceeds through an isocyanate intermediate [11]. Nevertheless, Colton et al. [12]

have established the presence of a chlorourea intermediate in the oxidation of urea by *t*-butyl hypochlorite under basic conditions,

$$NH_2CONH_2 + OCl^- \rightarrow NH_2CONHCl + OH^-, \tag{11.5}$$

strongly suggesting a Hoffman-type rearrangement:

$$NH_2CONHCl + OH^- \rightarrow H_2O + [NH_2CONCl]^- \rightarrow NH_2NCO + H_2O + Cl^-. \tag{11.6}$$

The hydrazine formed in Reaction 11.4 can be further oxidized to nitrogen by the hypochlorite:

$$2OCl^- + N_2H_4 \rightarrow N_2 + 2H_2O + 2Cl^-. \tag{11.7}$$

Hydrazine oxidation by hypochlorite is not as fast, however, as by chloramine [6] and, hence, a urea excess of only ~20% is required to yield hydrazine at ~62% based on hypochlorite, which is also a typical figure for the Hofmann degradation of amides. The oxidation of urea by *t*-butyl hypochlorite [12] produced much smaller yields (about 20%) at these urea/hypochlorite ratios, most likely due to the fact that the reaction did not follow an optimal time–temperature curve of the type shown in Figure 11.1.

In the absence of base, the reaction proceeds very vigorously with the release of carbon dioxide,

$$NH_2CONH_2 + 3OCl^- \rightarrow N_2 + 2H_2O + CO_2 + 3Cl^-, \tag{11.8}$$

which follows from combining Reaction 11.4 and Reaction 11.7, although the mechanism in acidic solution might not involve the formation of a hydrazine intermediate. In any case, no hydrazine can be detected in acidic media [12].

Reaction 11.7 is catalyzed by transition metal ions as is the Raschig process, although to a lesser degree [12]. In the present experiment, these are bound by the addition of a small amount of EDTA (~1 g/mol hypochlorite) [13]. The use of EDTA substantially reduces foaming and improves the purity of final product by keeping impurities in solution rather than in suspension. Because hydrazine is essentially an intermediate in both the oxidation of urea and in the Raschig process, yield is significantly dependent on the time–temperature curve for the reaction [6]. Thus, a reduction in total reaction time to ~2 h in the present experiment corresponds to ~30% decrease in yield. The thermal regime presented here is not universal, applying only to quantities of reagents used in the present experiment. Scaling-up is not straightforward, and some trial and error is required due to the exothermicity of the reaction.

Several methods exist for separating the hydrazine intermediate during the reaction, with formation of the ketazine by reaction with ketones/aldehydes, and the semisoluble bisulfate $N_2H_4.H_2SO_4$ (1 g/100 mL in 5% H_2SO_4 at 25°C) being the most common. The present preparation uses the latter method, which requires preliminary neutralization of the hydroxide present in equimolar amount in Reaction 11.4. Neutralization is

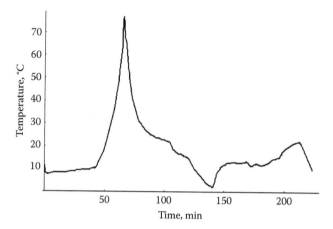

FIGURE 11.1   The time–temperature regime for the Hofmann degradation of urea. Slow mixing of the reagents below 10°C is followed by rapid heating and cooling to maximize the yield of the hydrazine intermediate. The final stage consists of an exothermic neutralization during which hydrazine precipitates as the sulfate.

performed with HCl rather than the more commonly used H_2SO_4 [7], reducing the exothermicity of the reaction and thermal loss of hydrazine, as well as co-precipitation of $Na_2SO_4.10H_2O$ on cooling, due to NaCl forming part of the salt load.

An analysis of gas evolution during Reaction 11.4 has shown that oxidation to nitrogen, Reaction 11.7, accounts for almost all yield loss in the thermal regime of Figure 11.1. Nitrogen evolved during the initial addition of hypochlorite and subsequent thermal spike accounts for 23% of yield loss, while another 8% is lost in the neutralization stage. The remaining 7% can be accounted for by hydrazine hydrate/carbonate volatilized with carbon dioxide in the neutralization stage as well as $N_2H_6SO_4$ dissolved in the ~5% H_2SO_4 solution at the end of the reaction.

11.2.2   HYDRAZINE HYDRATE

The literature preparation of hydrazine from its sulfate involves distillation from a stoichiometric amount of solid KOH, requiring considerable heat (~150°C), and therefore a copper or silver retort [8]. This is a low-yielding and cumbersome process due to the high hydration enthalpy of hydrazine 8.1 kJ/mol [14] and its low stability, yielding only 25% of the hydrazine as the monohydrate $N_2H_4.H_2O$. Distillation of hydrazine is also undesirable due to its considerable health effects even at low concentrations.

Presented here is a method based on the substantial solubility of hydrazine in the lower alcohols, so that ~74% hydrazine hydrate at yields of 65–75% can be obtained in a low-temperature extraction into alcohol after neutralization with NaOH/KOH as follows:

$$N_2H_6SO_4 + 2NaOH \rightarrow N_2H_4.2H_2O + Na_2SO_4. \tag{11.9}$$

Because each molecule of hydrazine is associated with two molecules of water, the concentration of hydrazine hydrate is 74% from stoichiometry. The alcohol can be removed by distillation or the solution can be used directly. Because commercial KOH contains about 10% bound water, the maximum achievable hydrazine hydrate concentration is 63% in that case.

11.2.3 HYPOCHLORITE TITRATION WITH IODINE

Hypochlorite ion solutions disproportionate to the chloride and chlorate ions on standing with the rate of decomposition being temperature and concentration dependent (as an example, a decomposition from 11.5% to 8.2% OCl⁻ over the course of 80 days at 30°C–40°C was measured by the author). Since from the previous section it is clear that the yield is sensitive to initial concentrations of reagents, it is important to establish accurately the hypochlorite concentration. This is most easily done iodometrically, adding excess KI to convert the hypochlorite to an equivalent amount of free iodine,

$$ClO^- + 2I^- + 2H^+ \rightarrow Cl^- + I_2 + H_2O, \tag{11.10}$$

and the I_2 titrated to a clear solution with thiosulfate and starch as usual:

$$I_2 + 2S_2O_3^{2-} \rightarrow 2I^- + S_4O_6^{2-}. \tag{11.11}$$

The following procedure was used to titrate the 11.5% NaOCl solution employed in this preparation: 5 mL of the reagent was diluted to 100 mL with distilled water; 5 mL of this diluted solution was placed in a conical flask, approximately 50 mL of water added, followed by 5 mL of 10% KI solution, and 1 mL glacial acetic acid. This was titrated with a solution of 2.5 g $Na_2S_2O_3$ made up to 100 mL. A few drops of concentrated starch suspension were added when the iodine color began to fade, and the hypochlorite strength is given by

$$(percent\ NaOCl) = 0.3725\ (volume\ Na_2S_2O_3\ in\ mL). \tag{11.12}$$

11.2.4 HYDRAZINE TITRATION WITH IODINE

The hydrazine yields quoted in this section were determined by means of iodiometric titration [4] according to the following reaction:

$$N_2H_4 + 2I_2 \rightarrow N_2 + 4HI. \tag{11.13}$$

Due to the basic nature of hydrazine, the HI produced by iodine reduction forms the $N_2H_4 \cdot HI$ adduct, which is not easily oxidized. For this reason, the titration is carried out with excess solid $NaHCO_3$ added to neutralize the HI. Typically, a 5% iodine solution (2.5 g iodine made up to 50 mL with methanol) is used, and the solution is vigorously stirred during the titration. Initially, the reaction is very rapid; however, after about half the iodine has been added, it slows down considerably due to the low solubility of $NaHCO_3$, so that several minutes of stirring are required to ensure an end point has been reached.

11.3   EXPERIMENTAL

11.3.1   HYDRAZINE SULFATE

It is not necessary to use costly analytical-grade urea in this preparation. The technical-grade urea suffices, provided it is treated with ammonia to a pH of ~10.5 subsequent to dissolution in water, followed by standing for several hours and filtration of precipitates of iron hydroxide, etc.

To begin, 76 g (1.9 mol) of NaOH is dissolved in 68 mL H_2O (this is about the minimum amount of water required at 25°C) and carefully added to 560 g (475 mL, 0.73 mol) of 11.5% NaOCl solution (these quantities can be adjusted if a hypochlorite solution of slightly different concentration is used), and the mixture cooled to approximately −10°C. Although the basified hypochlorite solution can be used at initial temperatures of up to 10°C, lower initial temperatures allow faster hypochlorite addition and, hence, shorter reaction times.

Next, 54 g (0.9 mol) of urea is dissolved in 46 mL of water (approximately the minimum amount). The dissolution is very endothermic and, hence, some heating is necessary to speed the process. The solution is poured into a 1-L, two-neck, flat-bottom reaction flask equipped with a thermometer and a pressure-equalized dropping funnel, and surrounded by an ice–salt mixture. The dropping funnel outlet is loosely blocked by glass wool to hinder the inflow of oxygen (which slowly oxidizes the hydrazine) and is either vented in the fume hood or bubbled through an acidified $KMnO_4$ solution that oxidizes the hydrazine.

A 3–4-cm stir bar is then inserted, 0.7 g EDTA is added, and the solution stirred while cooling to a temperature of about 8°C, at which point it becomes opaque and quite viscous. The basified hypochlorite solution is poured into the dropping funnel and run into the urea solution at such a rate that the temperature does not rise above 10°C. This should correspond to about 3–4 drops/sec., while the whole addition takes about 40 min. A much faster addition rate with a total addition time of 10 min raises the temperature to 25°C toward the end of the reaction with an increase in gas evolution due to Reaction 11.7, and yield loss of about 30%. However, even in this case, there is no substantive frothing.

When the reagents are fully mixed at a temperature of about 10°C, the solution is still noticeably yellow due to a substantial portion of unreacted hypochlorite. The temperature now needs to be raised to 72°C as quickly as possible to complete Reaction 11.4 and maximize the yield of the hydrazine intermediate, since a slow reaction favors the loss mechanism, Reaction 11.7. In practice, it is only necessary to heat externally to about 40°C, since a rapid exotherm (~4°C/min) commences at this point. As the temperature rises above 50°C, the deepening of color is due to the unreacted hypochlorite (Figure 11.2). When 72°C is reached, the reaction vessel is rapidly removed from the water bath and lowered into ice water. As the mixture cools below 50°C, it becomes essentially colorless, showing that all hypochlorite has reacted.

FIGURE 2.5 The nickel separator dipping below the molten NaOH encloses the cathode and forms the acid/base interface in the cell, and the region where hydrogen is generated. In Regime 2, this prevents sodium from shorting the separator to the cathode. The cathode consists of liquid sodium attracted by electrical forces to a water-cooled copper electrode raised just above the surface of the melt.

FIGURE 2.6 Sodium as it is collected from the cell and stored beneath liquid paraffin.

FIGURE 3.1   Molten KOH at 410°C electrolyzed in the same arrangement as NaOH with currents in the range 10–50 A, produces fleeting metallic streaks in the vicinity of the cathode. Fuming is greatly decreased under a protective argon stream, but no potassium globules are formed, nor is oxygen effervescence observed at the anode.

FIGURE 3.2   A 3-in.-long MgO crucible, which is not attacked by the caustic, is used to separate the anode and cathode compartments in the KOH cell. A water-cooled copper tube acts as the cathode and supports the crucible. Unlike the NaOH case, no hydrogen is generated in the cathode compartment; however, the copper–MgO juncture is not airtight to allow the fluid level inside the crucible to equilibrate.

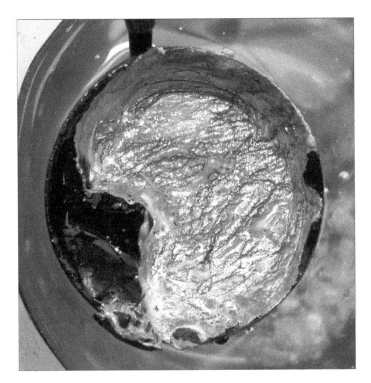

FIGURE 3.3 When the contents of the metal ladle have cooled under paraffin, the potassium is scraped from the solidified KOH layer below into the paraffin-filled container. The potassium has a blue-violet tinge due to traces of oxygen, and resembles soft butter in consistency.

FIGURE 4.1 A lithium globule on the surface of a molten KCl–LiCl eutectic at 420°C after electrolysis. At this temperature, lithium ordinarily burns in air; however, a thin film of electrolyte protects it. The film is made visible by traces of carbon from the anode adhering to the surface.

FIGURE 4.3 The molten lithium acts as the cathode being attracted to the steel electrode by electrical forces. This arrangement uses the surface tension of lithium to reduce its fragmentation, which otherwise leads to small lithium globules reaching the anode where their protective electrolyte coat is broken and an explosive reaction with chlorine results.

FIGURE 4.4 Lithium is collected from the cell with a perforated ladle. With the protective coat of electrolyte broken, surface oxidation commences and the molten lithium may ignite. It is thus quickly run into paraffin and held down with steel gauze.

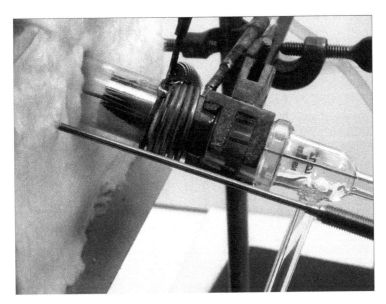

FIGURE 5.2 The evacuated lower end of the quartz reactor tube with side arm and detachable cesium receptor. Cesium condenses ahead of the region cooled by the copper tubing, coalesces into drops and flows into the receptor.

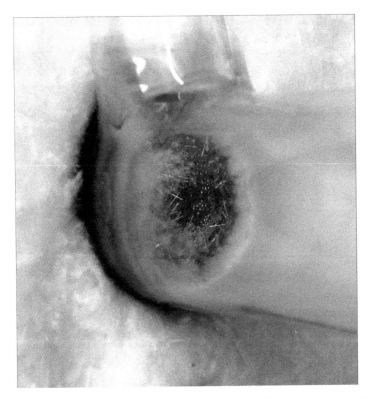

FIGURE 6.2 Sodium evaporating from an oven at 600°C, in hydrogen at 1 atm. NaH needles form in a section of the reactor where the temperature is sufficiently low for hydrogen pressure above NaH to be below 1 atm.

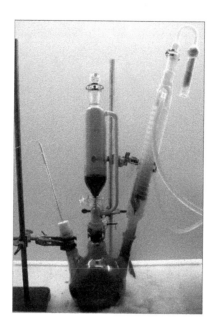

FIGURE 8.1 Bromine reacts with aluminum granules accompanied by sporadic flashing with the temperature rising to 210°C–220°C. The heat volatilizes the unreacted bromine, which refluxes in the condenser until it is all consumed. On cooling, the Al_2Br_6 product crystallizes as white scaly crystals on the flask walls.

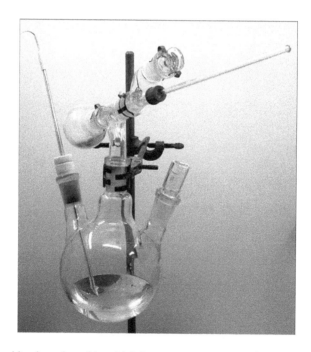

FIGURE 9.1 Aluminum bromide, which decomposes many organic solvents, is here slowly added to diethyl ether with the temperature maintained below 0°C by rapid swirling and cooling the ether in a freezing mixture. The glass stopper is replaced by a $CaCl_2$ protection tube prior to dissolution.

FIGURE 10.2 The reaction between sodium and Et$_2$BrAl commences at the melting point of sodium and becomes extremely vigorous at about 107°C. After a few minutes, the flask fills with large, black, porous pieces of Al–NaBr aggregate that soak up the Et$_3$Al product.

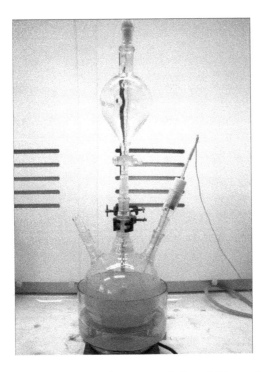

FIGURE 11.2 Some hypochlorite remains unreacted after addition to the cooled urea solution. Here, the temperature of the water bath is rapidly raised to 72°C to maximize the yield of hydrazine. About 8% hydrazine is lost at this stage, indicated by N$_2$ effervescence.

FIGURE 12.1 Diazotization of hydrazine with isopropyl nitrite in a basified alcoholic solution precipitates the insoluble azide salt. Potassium azide forms much larger crystals than sodium azide and is easier to purify.

FIGURE 13.2 Schlosser's base at −10°C metallates gaseous hydrogen with the formation of a white potassium hydride precipitate. The hydride is stable at room temperature; however, being pyrophoric in air, it is used in situ.

FIGURE 14.1 At 360°C sulfur in a three-neck flask burns in acetylene producing H_2S, CS_2, and a small amount of C_4H_4S (thiophene). The latter condenses in the ice water-cooled Liebig condenser and collects in a receiver. About 14% of the sulfur and traces of soot volatilized by the gaseous products also escape the reflux in the vertical section of the adaptor and are carried into the receiver with the CS_2.

FIGURE 16.4 Experimental setup for photolytic chlorination of chloroform. The chlorine generator, described in Chapter 15, is on the left in the photo. The 1-kW, high-pressure, mercury vapor lamp is on the right.

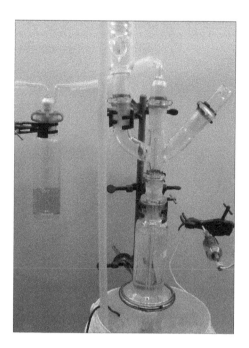

FIGURE 17.1 Setup for photolytical chlorination of dimethyl carbonate in carbon tetra-chloride. The free radical reaction is initiated by UV light from a 1-kW, Hg, high-pressure mercury vapor lamp. The bottle reactor is water cooled and the UV lamp is forced-air cooled. This arrangement increases the volume of liquid receiving useful radiation, as the absorption length for UV in the solution is very short.

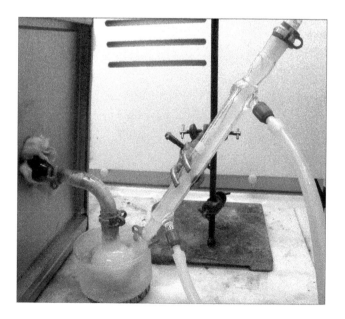

FIGURE 18.1 Calcium phosphide heated to 240°C reacts with a dull red glow with chlorine gas, yielding PCl_3, which is further oxidized to PCl_5. After about three quarters of the stoichiometric amount of chlorine has been introduced, PCl_5 sublimes out of the reaction zone and is deposited as light yellow flakes on the walls of the receiver flask.

FIGURE 19.1 A mixture of powdered carbon and $Ca_3(PO_4)_2$ is chlorinated at 760°C. Initially, $Ca_3(PO_4)_2$ is converted to $Ca(PO_3)_2$. Subsequently, $POCl_3$ forms and collects in the receiver. The gaseous products include CO and $COCl_2$; hence, the preparation is conducted inside a fume hood.

FIGURE 20.4 The fraction above 500°C from $NaHSO_4$ pyrolysis. Liquid oleum, colored black from carbonization of silicon grease, collects in the receiver below 580°C; above this temperature, nearly pure SO_3 deposits on the chilled flask walls by sublimation from the reactor.

FIGURE 21.1 Chlorine gas reacts instantly with sulfur at 160°C, producing S₂Cl₂ if sulfur is present in excess. The orange liquid S₂Cl₂ collects in the receiver; however, its color starts to deepen toward the end of the reaction as excess chlorine oxidizes the sulfur further to the deep red SCl₂.

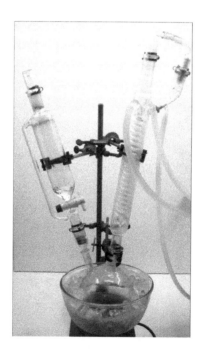

FIGURE 21.2 Thionyl chloride and chlorosulfonic acid are the major products when cooled sulfur dichloride is oxidized by sulfur trioxide, which is introduced here in the form of liquid oleum. Large amounts of sulfur dioxide are also produced, initially dissolving in the reaction mixture. The equilibrium is shifted to the right by heating. All tubing is polyethylene, as silicone is rapidly degraded by sulfur trioxide vapor.

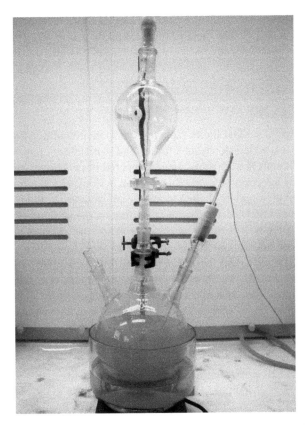

FIGURE 11.2 *A color version of this figure follows page 112.* Some hypochlorite remains unreacted after addition to the cooled urea solution. Here, the temperature of the water bath is rapidly raised to 72°C to maximize the yield of hydrazine. About 8% hydrazine is lost at this stage, indicated by N_2 effervescence.

When the temperature of the reaction mixture has dropped below 10°C, 188 mL of 32% HCl (1.9 mol) chilled to −10°C is run in while stirring at such a rate that the temperature does not exceed 10°C. This minimizes the decomposition of hydrazine (the nitrogen evolved at this stage accounts for about 8% yield loss) as well as hydrazine loss in the CO_2 flux. Initially, the addition is accompanied by the evolution of thick white fumes due to the formation of hydrazine hydrochloride; however, this soon clears. Next, a chilled mixture of 53 mL H_2SO_4 (1 mol) in 50 mL water is added at about 1 drop/sec., allowing the temperature to rise to just below 30°C. The initial stages of the addition are visually accompanied only by the evolution of residual CO_2 gas from the bicarbonate in solution because neutral hydrazine sulfate is very soluble. However, after about 0.3 mol of acid has been added, the bisulfate starts precipitating and the solution clouds. After all H_2SO_4 has been added, the solution is stirred for a further 15–30 min and then chilled in ice water. The added H_2SO_4 corresponds to a 0.55 molar excess, which lowers residual dissolved $N_2H_6SO_4$ to about 1 g. The time–temperature curve for the entire process is presented in Figure 11.1.

The precipitate is suction filtered, washed several times in situ with cold alcohol, and oven-dried at 100°C. The yield of the final product is 58–60 g $N_2H_6SO_4$, which is 62% based on hypochlorite.

11.3.2 Hydrazine Hydrate in Alcohol

11.3.2.1 Using KOH

To start, 114 g of 90% KOH flakes (about 1.8 mol) is ground to a fine powder in an electric grinder and added slowly, in a fume hood, with intermittent shaking to 101.5 g $N_2H_6SO_4$ (0.78 mol) in a 500-mL, glass-stoppered flask. The reaction generates its own moisture by Reaction 11.9, and the addition is slow, ~5 g at a time in the initial stages, with shaking regularly halted to let out any pressure buildup. Should caking occur, as evidenced by lack of substantial heating on the addition of KOH, the contents of the flask are well stirred with a glass rod and the addition continued. When all KOH has been added, 120 mL of absolute ethanol is poured into the flask, the contents are well shaken, and stirring of the now fluid mixture in the stoppered flask is continued magnetically for about 30 min. The precipitate of K_2SO_4 is filtered and washed together with the reaction vessel several times with fresh absolute ethanol until 240 mL of the latter has been added. About 73% of the hydrazine is recovered in solution. Filtration under gravity is fairly slow and can be accelerated by aspirator vacuum; however, excessively rapid filtration reduces the yield by about 10%.

Although dissolution of KOH in the alcohol prior to reaction with hydrazinium sulfate appears more convenient since the reaction now involves adding liquid to a solid, it produces a substantial amount of insoluble gel due to the insolubility of the sulfates in alcohol, which lock a substantial portion of the hydrazinium sulfate and so reduce the yield.

11.3.2.2 Using NaOH

This produces hydrazine hydrate with 26% water content, as opposed to 37% with KOH, since commercial NaOH contains less than 1% water. The procedure is the same as with KOH, with 73 g NaOH used. Due to the lower water content, the mixture tends to cake more and more frequent stirring with a spatula is required. The yield of hydrazine in solution is 71%, almost the same as with KOH.

The hydrazine solution should not be stored for lengthy periods in glassware, as both it and the hydroxides slowly attack glass resulting in a precipitate of sodium silicate generated over time.

Hydrazine is a suspect carcinogen with the permissible concentration level PCL of 0.01 ppm close to its detection threshold, and much below the average odor detection threshold of 2–3 ppm. If the odor of hydrazine is detected—it has a "fishy" odor resembling low bp aliphatic amines, and quite different from that of ammonia—it is being inhaled in above PCL levels.

REFERENCES

1. Sandler, S. R. and Karo, W., *Organic Functional Group Preparations*, p. 74, pp. 332–3, 341, 367–84. New York–London: Academic Press, 1968.

2. Roden, B. A., Hydrazine. In *Encyclopedia of Reagents for Organic Synthesis*, edited by Paquette, L. New York: John Wiley & Sons, 2004.

3. Lindsay, R. O. and Allen, C. F. H., Phenyl azide. *Org. Synth. Coll.* 3: 710, 1955.

4. Rothgery, E. F., Hydrazine and its derivatives. In *Kirk-Othmer Encyclopedia of Chemical Technology, 5th ed.*, Vol., 13. New York: John Wiley & Sons, 2006.

5. Furniss, B. S., Hannaford, A. J., Smith, P. W. G., and Tatchell, A. R., *Vogel's Textbook of Practical Organic Chemistry, 5th ed.*, pp. 960–1, 1149–50. London: Addison Wesley Longman, 1989.

6. Cahn, J. W. and Powell, R. E., The Raschig synthesis of hydrazine. *J. Am. Chem. Soc.* 76(9): 2565–67, 1954.

7. Audrieth, L. F., Nickles, T. T., Gibson, G., and Kirk, R. E., Hydrazine sulfate. *Inorg. Synth.* 1: 90–1, 1939.

8. Schenk, P. W., Nitrogen. In *Handbook of Preparative Inorganic Chemistry, 2nd ed.*, pp. 468–72, edited by Brauer, G. New York–London: Academic Press, 1963.

9. Audrieth, L. F. and Ogg, R. A. *The Chemistry of Hydrazine.* New York: John Wiley & Sons, 1951.

10. Schestakow, P., Verfahren zur Darstellung von Hydrazin und Seinen Derivaten. DE. Patent No. 164755, Feb. 7, 1903.

11. Hayashi, H., Hydrazine synthesis: Commercial routes, catalysis and intermediates. *Res. Chem. Intermed.* 24(2): 183–96, 1998.

12. Colton, E., Jones, M. M., and Audrieth, L. F., The preparation of hydrazine from urea and *t*-butyl hypochlorite. *J. Am. Chem. Soc.* 76(9): 2572–74, 1954.

13. Pfeiffer, P. and Simons, H., Beitrag zur Darstellung des Hydrazins. *Chem. Ber.* 80(2): 127–8, 1946.

14. Hughes, A. M., Corruccini, R. J., and Gilbert, E. C., Studies on hydrazine: The heat of formation of hydrazine and of hydrazine hydrate. *J. Am. Chem. Soc.* 61(10): 2639–42, 1939.

12 Sodium and Potassium Azide

SUMMARY

- Diazotization of alcoholic hydrazine hydrate yields 86% KN_3 and 90% NaN_3.
- The preparation avoids use of the toxic and unstable hydrazoic acid.
- KN_3 purity is >99%, NaN_3 >98% by IC analysis, nitrite is the major impurity.
- Facile preparation of isopropyl nitrite at 87% yield is presented.

APPLICATIONS [1]

- Curtius rearrangement of acyl chlorides to isocyanates.
- Schmidt reaction of carbonyls forming substituted amines.
- Staudinger reduction to amines by nucleophilic addition of substituted phosphines.
- Azide-alkyne Huisgen cycloaddition yielding substituted triazoles and its polymerization variant in so-called click chemistry.
- Synthesis of heavy metal azides used in primary explosives, etc. [2].
- Synthesis of high-purity alkali and alkaline earth metals [2].

12.1 INTRODUCTION

Sodium azide is usually readily available; however, potassium azide is sometimes required due to its enhanced solubility, especially in aprotic solvents such as DMSO [3], or greater thermal stability (mp KN_3 350°C, while sodium azide decomposes just above 275°C). Following is a very simple means for quickly preparing both azides by diazotization of hydrazine hydrate in alcohol, avoiding work with the highly toxic and explosive hydrazoic acid. The yield is about 86% for potassium azide and over 90% for sodium azide based on hydrazine (the former yield is smaller due to the ~10% water content in commercial potassium hydroxide, which retains some of the azide in solution). The potassium azide precipitates more slowly than sodium azide and in larger crystals and, hence, is easier to purify.

12.2 DISCUSSION

Diazotization was discovered in 1858 by Peter Griess who prepared phenyl azide by diazotizing phenylhydrazine with nitrous acid. Since then, other aromatic hydrazines, anilines, alkyl amines, and hydrazines have been diazotized with the stability of the product depending on the effectiveness of the organic radical as a charge stabilizer. Hydrazine, the simplest form of the latter class of compounds, was diazotized by Curtius in 1890, thus preparing an aqueous solution of hydrazoic acid, HN_3, for the first time and providing an example of an inorganic diazotization reaction:

$$N_2H_4 + HNO_2 \rightarrow HN_3 + 2H_2O. \tag{12.1}$$

Hydrazoic acid can be isolated by distillation, an undesirable procedure because its toxicity is comparable to the cyanides, while it is also highly unstable and explosive. Distillation is also required if azide salts are the end products from Reaction 12.1, because the nitrous acid is generated in situ by the reaction between a nitrite and a mineral acid, thus introducing foreign ions into solution.

The present preparation avoids generating free hydrazoic acid entirely by carrying out Reaction 12.1 in an alcoholic medium in which hydrazine hydrate is sufficiently soluble, and the alkali azides are insoluble (solubility of NaN_3 in alcohol 0.5 g/100 mL at 20°C) and so precipitate out. The alcohol is preliminarily esterified with HNO_2 to yield the corresponding organic nitrite, which is soluble in alcohol:

$$ROH + HNO_2 \rightarrow RNO_2 + H_2O. \tag{12.2}$$

The nitrite is then used to diazotize hydrazine in a basic alcoholic solution from which the azide precipitates:

$$N_2H_4.H_2O + RNO_2 + NaOH \rightarrow NaN_3 + 3H_2O + ROH. \tag{12.3}$$

The choice of alcohol for the reaction medium is a compromise between the increased solubility of hydrazine hydrate in the lower alcohols and the volatility of corresponding nitrite ester. Thus, while solubility is highest in methanol, methyl nitrite is a gas with a boiling point of −17°C, while the solubility of hydrazine hydrate in isopropyl alcohol is relatively poor. For this reason, ethanol is chosen as a suitable compromise with ethyl nitrite (bp 17°C) being easily condensable in an ice water-cooled reflux condenser, while the solubility of hydrazine hydrate in ethanol is greater than 2.5 mol/L, from Chapter 11.

Although ethanol is the most convenient reaction medium, esterification in Reaction 12.2 is more easily performed with isopropyl or n-butyl alcohols because their nitrite esters are sufficiently nonvolatile to be easily formed in good yield. Consideration of the volatility of ethyl nitrite is still relevant due to the following transesterification reaction involving metathesis of the nitrite:

$$BuNO_2 + EtOH \rightarrow BuOH + EtNO_2. \tag{12.4}$$

Hence, an efficient reflux condenser is arranged to prevent loss of the ethyl nitrite ester formed by transesterification.

12.3   EXPERIMENTAL

Nitrite esters decompose on standing over a period of weeks and, therefore, need to be prepared prior to use. There is no noticeable difference in yield between *n*-butyl alcohol and isopropyl alcohol nitrite esters, following the literature preparations for butyl nitrite [4] and isopropyl nitrite [5]. Both preparations are rather labor intensive and, hence, a new method is presented below that can be completed in about 15 min, gives a yield of 90%, and produces a sufficiently pure product for the present experiment.

12.3.1   ISOPROPYL NITRITE

First, 93 mL of 32% w/w HCl is mixed with 49 g isopropyl alcohol and cooled in a salt–ice bath. (isopropyl alcohol is insoluble in concentrated sodium nitrite solution; hence, the reagents must be mixed in this order). Then, 45 g of sodium nitrite (0.65 mol) is dissolved with magnetic stirring in 80 mL water in a 500-mL flat-bottom flask. This dissolution is substantially endothermic; thus, if done immediately prior to the reaction, no external cooling is required. The HCl–isopropanol mixture is placed into a suitable dropping funnel with vented outlet and added to the vigorously stirred sodium nitrite solution in a steady stream over the course of 10–15 min, during which a small amount of NO_2 gas may form. Next, the same dropping funnel is used to separate the lower aqueous layer from the yellow oily isopropyl nitrite, which is then shaken with a similar volume of concentrated cooled sodium bicarbonate solution (shaking at room temperature results in a sudden pressure buildup). One obtains 50.4 g (0.57 mol) of isopropyl nitrite, which completely distills at 38°C–39°C, representing an 87% yield.

12.3.2   POTASSIUM AZIDE

Begin by pouring 230-mL of absolute ethanol into a 500-mL, flat-bottom, single-neck flask placed on a heater/stirrer and then 19 g (0.38 mol) hydrazine hydrate is added. Alternatively, one can use 230 mL of ethanolic hydrazine hydrate solution prepared in Chapter 11 using KOH as the base. Next, 26.6 g (0.4 mol) of 85% KOH is added to the flask and dissolved with magnetic stirring. A dropping funnel with pressure equalization is now attached, and 39 g (0.43 mol) of isopropyl nitrite (alternatively, 44.7 g *n*-butyl nitrite) is placed in the funnel. A double-surface reflux condenser is attached to the dropping funnel and a steady stream of ice water is passed through the coil.

The isopropyl nitrite is now run into the basified alcoholic hydrazine hydrate solution at about 2 drops/sec. (faster initially) so as to maintain a steady reflux of colorless ethyl nitrite in the reflux condenser (Figure 12.1). The addition is at first accompanied by the appearance of white KN_3 fumes that fill the entire apparatus. This is due to the substantial vapor pressure of isopropyl nitrite and hydrazine, so that part of the reaction takes place in the vapor phase; however, this soon disperses and the vapor pressure of hydrazine drops. The level of nitrite ester in the funnel gradually decreases as it is consumed by the reaction. The advantage of the present arrangement is that should an excessive amount of the nitrite be added, it is immediately refluxed back into the dropping funnel.

FIGURE 12.1 *A color version of this figure follows page 112.* Diazotization of hydrazine with isopropyl nitrite in a basified alcoholic solution precipitates the insoluble azide salt. Potassium azide forms much larger crystals than sodium azide and is easier to purify.

After about 30 min, there is no isopropyl nitrite left in the dropping funnel and the hotplate is heated to 100°C so as to maintain a steady reflux of ethanol/isopropanol for 10–15 min (this step improves the yield by a few percent). The reaction flask, whose contents are now almost colorless, is placed in an ice-cold bath, and the potassium azide left to crystallize over 1–2 h. White bulky crystals of flat polygonal shape soon crystallize over the entire volume of the flask. They are separated from the solution by filtration, and the flask, as well as precipitate, washed several times in cold ethanol followed by cold ether. The crystals, which should be shining white, yield 26.5 g KN$_3$ or 86% based on hydrazine.

12.3.3 SODIUM AZIDE

The procedure here is similar to that used in the previous subsection, with 16.2 g 99% NaOH used in place of KOH to basify the ethanol. If the hydrazine hydrate is prepared as per the hydrazine section, NaOH rather than KOH should be used to basify the solution avoiding contamination with potassium ions. Sodium hydroxide dissolves with considerably greater difficulty in ethanol than potassium hydroxide, producing a cloudy solution. The yield of sodium azide after washing twice in cold ethanol and ether is 93%; however, it is significantly harder to purify than potassium azide because it crystallizes rapidly in a semiamorphous mass. After washing many times with cold alcohol and ether, an essentially white powder can be obtained at a yield of 89%.

REFERENCES

1. Bräse, S., Gil, C., Knepper, K., and Zimmermann, V., Organic azides: An exploding diversity of a unique class of compounds. *Angew. Chem. Int. Ed.* 44: 5188–240, 2005.
2. Ehrlich, P., Alkaline Earth metals. In *Handbook of Preparative Inorganic Chemistry, 2nd ed.*, pp. 927–8, edited by Brauer, G. New York–London: Academic Press, 1963.
3. Insalaco, M. A., and Tarbell, D. S., *t*-Butyl Azidoformate. *Org. Synth. Coll.* 6: 207, 1988.
4. Furniss, B. S., Hannaford, A. J., Smith, P. W. G., and Tatchell, A. R., *Vogel's Textbook of Practical Organic Chemistry, 5th ed.*, p. 414. London: Addison Wesley Longman, 1989.
5. Levin, N., and Hartung, W. H., ω-Chloroisonitrosoacetophenone. *Org. Synth. Coll.* 3: 191, 1955.

13 Potassium *t*-Butoxide and Potassium Hydride

SUMMARY

- Hydrogenation of Schlosser's base at 1 atm with TMEDA catalyst yields active KH.
- Relatively high reaction temperature of −10°C to −5°C is achievable with an ice–salt mixture.
- KH metallates instantly and in situ the sterically hindered ketone pinacolone, allowing the reaction to be performed as a titration.
- The combined yield of Schlosser's base hydrogenation and pinacolone metallation is 72%.
- Enolate quenched with $Si(CH_3)_3Cl$ shows no elimination products in GC/ MS spectra.
- Active LiH is formed in similar fashion at room temperature in 76% yield.
- LiH does not metallate pinacolone, indicating the increased activity of KH.
- *t*-BuOK is prepared in quantitative yield and essentially free of alcohol (<1%).

APPLICATIONS

Potassium *t*-Butoxide [1]
- Alpha-alkylation of ketones, amides, and esters.
- Acylation of amides and ureas.
- Promoter of aldol condensation.
- Michael addition and Stobbe condensation in difficult cases.
- Replaces aluminum alkoxides in the Oppenauer oxidation of alcohols to low-bp aldehydes particularly when substrate contains basic nitrogen [2].
- Synthesis of acetylenes in double-elimination reactions.
- In conjunction with alkyl-lithium compounds forms Schlosser's superbase.

Potassium Hydride [3,4]
- Order of magnitude more reactive than NaH, metallates alkenes, amines, sulfoxides, etc.
- Limited reactivity toward nucleophilic addition [4,5].
- Metallation products used in usual way to form ethers or extend carbon skeleton.

13.1 INTRODUCTION

These reagents became popular several decades ago as very strong bases in aprotic media, such as DMSO, hexanes, and THF [5], but showing limited activity toward nucleophilic addition, reduction, or coupling reactions [1,3–5].

Potassium *t*-butoxide, being a stronger base than primary or secondary alkoxides (as *t*-BuOH is a weaker acid), is useful as an intermediate in a multitude of reactions [1]. Moreover, in a 1:1 molar mixture with alkyl-lithium compounds, such as *n*-butyl-lithium, it forms the so-called Schlosser's superbase, which is capable of alkylating (metallation followed by electrophilic substitution) very weak acids, such as alkenes, amines, sulfoxides, and ketones [5].

On exposure to the atmosphere, potassium *t*-butoxide is decomposed to KOH and K_2CO_3, which can decrease the concentration of a frequently used stock to as little as 30% in a year [1]. It thus is often either prepared in situ by reacting potassium with *t*-butyl alcohol [6,7] or purified by sublimation prior to use. The nominal basic strength of *t*-BuOK, as measured by its reactivity toward weak acids, is dependent on the pK_a of the solvent, decreasing in the order DMSO > alkanes > THF > *t*-BuOH. A difference in basicity also arises from the formation of butoxide clusters [*t*-BuO-]$_m$[Solvent]$_n$ both in the solid state (explaining its volatility) and in solution, with the clusters being less reactive with the proton [1]. Hence, in applications such as Schlosser's superbase, this reagent must be prepared free of the parent alcohol, and this poses the difficulty of separating the alcohol from the very basic alkoxide. A literature method for preparing *t*-BuOK free from alcohol [7] uses a large excess of *t*-BuOH (28 mL *t*-BuOH per gram K), and substantial sublimation of the product occurs during the vacuum separation, blocking the apparatus. A modified procedure is presented here using 18 mL *t*-BuOH per gram K, which shows no observable sublimation of the product through careful temperature control.

Metallation with hydrides as opposed to organoalkali compounds has the advantage that gaseous hydrogen is the only by-product. Their chief drawback has been their insolubility in organic solvents that limits reactivity. As mentioned in Chapter 6, the stability of these ionic compounds decreases and, thus, reactivity increases with increasing size of the metal cation. Thus, LiH forms by a direct combination of the elements at 700°C, while NaH requires higher pressures for a substantial reaction to occur. It is interesting to note that Klusener et al. [3] report that KH precipitates in high yield by metallating hydrogen at atmospheric pressure with Schlosser's base in hexane at −25°C–20°C. The KH is easily separated from the reaction mixture by the dissolution of *t*-BuOLi in benzene/toluene followed by filtration [8]. KH can be a more convenient metallating agent than Schlosser's base (*n*-BuK) since the evolution of hydrogen gas (butane is liquid at typical metallation temperatures) gives a clear indication of the progress of the reaction and avoids the use of excess reagent, which might cause problems during workup.

Potassium hydride (a much stronger base than *t*-BuOK [9]) is an order of magnitude more reactive than the commercially available NaH (with commercial LiH even less reactive) capable of metallating very weak acids, such as alkenes, amines, sulfoxides, and ketones, at room temperature and showing limited reducing activity toward nucleophilic addition [3,4]. Moreover, the hydrides formed from Schlosser's reagent are more reactive by several orders compared to those obtained by thermal methods (to the extent that they are pyrophoric in air). Thus, KH metallates very weakly acidic and sterically hindered ketones, such as *t*-butyl methyl ketone (pinacolone) instantly at −20°C, while similarly prepared NaH and LiH require ~20 min, with the latter producing a substantial amount of addition product, evident after quenching with trimethylchlorosilane. In contrast, commercial NaH and LiH react to the extent of only 20% and 2%, respectively [3].

The present synthesis yields KH at 85–90% based on Schlosser's reagent and is a modified version of the method of Klusener et al. [3], proceeding at higher temperatures of −10°C–5°C, which can be conveniently achieved with an ice–salt freezing mixture. The KH is not isolated but used in situ; hence, the reported yield is based on the active hydrogen content, in this case established by metallation of pinacolone. This reaction is carried out at −10°C to minimize elimination and double metallation reactions and is instantaneous, so that addition of the ketone can be performed as a titration. The enolate so formed is quenched with methyl iodide and trimethylchlorosilane ($Si(CH_3)_3Cl$) to give the corresponding ethers in good yield as indicated by GC/MS spectra.

Active LiH was also prepared by hydrogenating BuLi/TMEDA in the absence of *t*-BuOK, demonstrating that the small amount of LiH present in KH due to incomplete transmetallation of Schlosser's base is essentially inactive toward pinacolone. It is found capable of metallating relatively acidic ketones, such as acetone ($pK_a = 19$), instantly, but is totally inert toward less acidic ketones, such as pinacolone ($pK_a \sim 28$ [10]), even at 20°C–30°C.

13.2 DISCUSSION

13.2.1 THE RLI–TMEDA COMPLEX

Although the Li-C bond is predominantly ionic (55–95%) in character, alkyllithium compounds show high solubility in organic solvents due to the formation of clusters with Li and R-groups occupying alternate tetrahedral or octahedral vertices [11]. Thus, *n*-BuLi forms a hexameric Li_6 cluster in hexane, whose covalent character is dictated by the bonding in the large alkyl groups surrounding the core of ionic Li-C bonds with the former completely determining the physical properties of the organometallic.

The steric hindrance posed by this clustering also serves to reduce the chemical reactivity of the Li-C bond. The presence of tertiary amines (lacking active hydrogens) greatly increases the reactivity of alkyllithium compounds [5]. This increase in reactivity is not a catalytic effect, but rather due to the formation of the thermodynamically more stable Li-N coordination bonds that break up the Li cluster, as evidenced by the fact that greatest reactivity increase is observed with substantial R_3N-to-RLi molar ratios. Among the most effective activators of alkyllithium compounds is tetramethylethylenediamine (TMEDA), as the bidentate coordination bonds result in an especially stable Li/TMEDA chelate, which in solution is effectively a monomer [12].

n-BuLi is unstable in diethyl ether and tetrahydrofuran (THF) with cleavage of the ether link occurring on time-scales of the order of several hours [13]. It is relatively stable in hydrocarbons with a ~2 M solution in hexane decomposing at a rate <0.01% per day at 25°C by reduction to butene:

$$C_4H_9Li \rightarrow C_4H_8 + LiH. \tag{13.1}$$

However, the decomposition rate rises to 0.05% per day and 13% per hour at 45°C and 130°C, respectively. The presence of moisture or carbon dioxide leads to further decomposition:

$$C_4H_9Li + H_2O \rightarrow C_4H_9OH + LiOH. \tag{13.2}$$

Therefore, stock solutions of BuLi should be titrated before use. While many methods are available [12], unlocked nuclear magnetic resonance (NMR) provides an interesting nonwet alternative [14]. The instrument is preliminarily locked and shimmed with a deuterium signal, and subsequently run in unlocked mode (as $CDCl_3$ and n-BuLi react). The author has found some variability using benzene as an internal concentration standard in old stock n-BuLi, likely due to mixing difficulty; however, the hexane solvent itself can be used as a reference standard, in a method that is simpler and less subject to variation. The procedure is to form a ratio between the $LiCH_2$ proton integral at −1.0 ppm (*a*), and the hexane proton signal measured as the integral above +1 ppm (*b*). Using the molecular weights of hexane and n-BuLi and the essentially constant density of 0–3 M n-BuLi solutions in hexane, 680 g/L at 20°C, the following expression has been found accurate to a few percent:

$$\text{Molarity n-BuLi} = \frac{55a}{b + 1.7a}. \tag{13.3}$$

13.2.2 Schlosser's Superbase

Treatment of an n-BuLi/TMEDA solution in hexane with a suspension of n-BuOK in the same solvent results in a reaction evidenced by a color change to yellow at −20°C (brown at −5°C). This is accompanied by the formation of n-BuK in the following transmetallation reaction:

$$n\text{-BuLi} + t\text{-BuOK} \rightarrow n\text{-BuK} + t\text{-BuOLi}. \tag{13.4}$$

Although the equilibrium lies to the right [15], the reaction rate is fairly slow as is evident from the fact that only a small portion of the t-BuOK suspended in hexane dissolves. The reaction rate can be increased significantly by adding benzene or toluene in which t-BuOLi is substantially soluble [16]. The same effect is achieved by using the superbase in situ to metallate weakly acidic compounds (e.g., alkenes or sterically hindered ketones, such as pinacolone), which are not metallated directly by n-BuLi/TMEDA to a significant extent. The superbase also metallates gaseous hydrogen [3] according to the reaction

$$n\text{-BuK} + H_2 \rightarrow n\text{-BuH} + KH, \tag{13.5}$$

with yields of ~80% obtained here. However, pure *n*-BuK is less selective toward olefinic metallation than Schlosser's base [5]; hence, representation of the latter by *n*-BuK is one of convenience.

Metallation rates of weakly active hydrogen compounds, such as allylic methyl groups or enolized ketones,

$$(CH_3)_3COCH_3 + KH \rightarrow (CH_3)_3C(OK)CH_2 + H_2, \tag{13.6}$$

greatly exceed addition reactions rates

$$(CH_3)_3COCH_3 + KH \rightarrow (CH_3)_3CH(OK)CH_3, \tag{13.7}$$

while double metallation only proceeds when the base is present in substantial excess [17].

The metallated products can be used in the usual way, for instance, quenched with trimethylchlorosilane or methyl iodide to form ethers,

$$ROK + CH_3I \rightarrow ROCH_3 + KI, \tag{13.8}$$

or reacted to extend the carbon skeleton, for instance, with epoxides:

$$RK + (CH_2)_2O + H_2O \rightarrow RCH_2CH_2OH + KOH. \tag{13.9}$$

13.3 EXPERIMENTAL

The *n*-BuLi used in these preparations is readily synthesized from *n*-butyl alcohol, sodium bromide, and lithium metal as described in the literature [6,18].

13.3.1 POTASSIUM *T*-BUTOXIDE

185 mL of *t*-BuOH is placed in a 1-L, two-neck flask in an oil bath and fitted with an efficient reflux condenser. Then, 10.5 g of potassium is shaken with about 100 mL dry hexane to dissolve adhering oil and, while still under hexane, cut with the edge of a spatula into about 20 pieces. Dry nitrogen is admitted into the flask through one of the necks and, as each piece is cut, it is dropped into the alcohol through the flask neck serving as the nitrogen outlet; this minimizes exposure of the potassium to the atmosphere. After all the potassium has been added, the bath temperature is raised to establish a gentle reflux in the condenser (about 90°C–100°C). Hydrogen evolution is very measured under these conditions and slows considerably toward the end, so that approximately 2 h is required for the potassium to react completely. The latter is greatly accelerated by localized superheating at the K–*t*-BuOH interface due to the heat of the reaction, so that magnetic stirring does not increase the reaction rate, but rather can decrease it as it reduces the superheating.

The setup is now rearranged for downward distillation through a Liebig condenser (more efficient condensers block, as the mp of *t*-BuOH is 26°C) into a 250-mL receiver, and excess *t*-BuOH is distilled off to a final bath temperature of 160°C. Although the bp of *t*-BuOH is 78°C, at this point only about half the excess alcohol has distilled off due to the formation of alkoxide/alcohol clusters (see section 13.1), and the mixture is still fluid. Further substantial evaporation of alcohol at atmospheric pressure can only be effected at the expense of substantial decomposition; hence, aspirator vacuum (12–20 torr) is applied to the receiver outlet through a calcium chloride drying tube, after the distillation flask has cooled to below 60°C (or else the flask contents, may be ejected into the condenser due to excessive vigorous boiling). As additional alcohol distills off at reduced pressure, the bath temperature is raised to 180°C. The temperature in the distilling flask should not be allowed to drop during this procedure, as this can lead to premature solidification of the flask contents, which significantly slows down subsequent drying (premature solidification cannot be reversed by raising the temperature again). The resulting white powder is allowed to cool below 60°C, the distilling flask connected to a liquid nitrogen trap (after the receiver has been removed to prevent the *t*-BuOH content subliming into the LN$_2$ trap), and finally dried in an oil pump vacuum (0.5–1 torr). Over the course of several hours, the bath temperature is raised to 130°C, but should not be allowed to exceed 140°C, because *t*-BuOK begins to sublime at this point, coating the entire apparatus with a fine pyrophoric layer, which is difficult to clean. The final yield of *t*-BuOK is 30.5 g (Figure 13.1); hence, about 0.3 g *t*-BuOH (1%) is still present in the product, which, however, is suitable for most purposes.

13.3.2 POTASSIUM HYDRIDE

The temperature of the freezing bath is monitored throughout the procedure and is not allowed to rise above −5°C. A 100-mL, single-neck flask equipped with a septum is flushed with dry nitrogen, and 30 mL of 2.1 M solution (determined as in the following text) of *n*-BuLi in hexane is injected. Next, 9.5 mL of TMEDA is added (a small amount of moisture in this is admissible as it is dried by the *n*-BuLi leaving a LiOH gel that does not transfer to the reaction flask), and the flask is cooled at −10°C (1:3 salt–ice mixture). A three-neck flask equipped with gas bubbler inlet, bubbler outlet, and septum, and a 4-cm magnetic stirrer is flushed with dry nitrogen. Then, 7 g *t*-BuOK is placed in the flask, 50 mL of hexane (dried by distillation from LiAlH$_4$) is added, and the flask contents cooled at −10°C. The *n*-BuLi–TMEDA mixture is transferred into the three-neck flask either by cannula or rapidly by syringe. The superbase forms immediately upon mixing as evidenced by a yellow/light brown coloration, and heat is evolved so that additional freezing of the mixture may be required.

With rapid stirring (*t*-BuOK is only very slightly soluble in hexane), hydrogen from a cylinder is admitted through a small P$_2$O$_5$ drying tube so that 1–2 bubbles/sec. pass the outlet bubbler. After a few minutes, the solution begins to lighten and a white precipitate separates (Figure 13.2). After 30–40 min, the reaction is complete (excess hydrogen is used) and is allowed to reach room temperature in a

FIGURE 13.1 Potassium *t*-butoxide, formed by the reaction of potassium metal with excess *t*-butanol, exists in the solid state as clusters held together by weak bridge bonds. It, therefore, sublimes at relatively low temperatures and separation from the parent alcohol, which is needed to increase its pH, must be performed under carefully controlled conditions.

hydrogen flux. Because KH is pyrophoric, it is metallated in situ by injecting pinacolone through the septum. Hydrogen is instantaneously released to a total of 1115 mL, corresponding to 0.045 mol, or a yield of 72% for the combined hydrogenation/metallation reaction. Mass spectra of the reaction products quenched with MeI and Me$_3$SiCl gave the corresponding allyl ethers with a significant amount of dimetallation evident in the MeI quench.

The strength of stock BuLi solutions is often determined by Gilman double titration [19], which evaluates the total base concentration of the hydrolyzed solution and subtracts the LiH/LiOH content. The latter is measured by quenching BuLi with an alkyl bromide and titrating (some sources suggest benzyl chloride [13], but this reaction is excessively vigorous at room temperature after an initial quiescent period of 4–5 min). For the present synthesis, it is most convenient to use quantitative NMR as outlined earlier or hydrolyze about 10 mL of n-BuLi stock solution by slow injection of excess 4:1 hexane/BuOH v/v and determine the molarity from the amount of hydrogen evolved. This method is insensitive to LiOH content and is typically accurate to a few percent.

FIGURE 13.2 *A color version of this figure follows page 112.* Schlosser's base at −10°C metallates gaseous hydrogen with the formation of a white potassium hydride precipitate. The hydride is stable at room temperature; however, being pyrophoric in air, it is used in situ.

13.3.3 Lithium Hydride

A more active form of LiH than that prepared in Chapter 6 can be synthesized at 30°C–35°C in a procedure that is otherwise similar to that for KH. In this case, due to the substantial volatility of hexane in a hydrogen flux, a Liebig condenser cooled with ice water is placed between the gas outlet and the bubbler.

Next, 20 mL of 2.1 M *n*-BuLi in hexane is injected into a nitrogen-flushed, three-neck, 100-mL flask placed on a heater/stirrer. Then, 60 mL hexane and 6.4 mL TMEDA are added, and the flask is heated on a bath to 30°C. Dry hydrogen is admitted so that about 1–2 bubbles/sec. pass through the bubbler. During this period, the reaction evolves sufficient heat to sustain itself, and the temperature rises to about 35°C. After 40 min, hydrogen absorption ceases and the product assumes the form of a white powder in a light-yellow liquid. Unlike the KH case, metallation with pinacolone at 30°C produces no signs of reaction, while acetone is metallated instantly, releasing 800 mL or 0.032 M gas, corresponding to a 76% yield for the combined hydrogenation/metallation reactions.

REFERENCES

1. Pearson, D. E. and Buehler, C. A., Potassium *t*-Butoxide. *Chem. Rev.* 74(1): 45–86, 1973.
2. Woodward, R. B., Wendler, N. L., and Brutschy, F. J., Quininone. *J. Am. Chem. Soc.* 67(9): 1425–9, 1945.
3. Klusener, P. A. A., Brandsma, L., Verkruijsse, H. D., von Ragué Schleyer, P., Friedl, T., and Pi, R., Superactive alkali metal hydride metalation reagents: LiH, NaH, and KH. *Angew. Chem. Int. Ed. Engl.* 25(5): 465–66, 1986.
4. Brown, C. A., Potassium hydride: A highly active new hydride reagent. Reactivity, applications, and techniques in organic and organometallic reactions. *J. Org. Chem.* 39(26): 3913–18, 1974.
5. Schlosser, M., Superbases for organic synthesis. *Pure Appl. Chem.* 60(11): 1627–34, 1988.
6. Furniss, B. S., Hannaford, A. J., Smith, P. W. G., and Tatchell, A. R., *Vogel's Textbook of Practical Organic Chemistry, 5th ed.,* p. 744. London: Addison Wesley Longman, 1989.
7. Skattebøl, L. and Solomon, S., *Org. Synth. Coll.* 5: 306–9, 1973.
8. Schlosser, M., Prescriptions and ingredients for controlled CC bond formation with organometallic reagents. *Angew. Chem. Internat. Engl. Ed.* 13(11): 701–6, 1974.
9. Brown, C. A., The remarkable fast reaction of potassium hydride with amines and other feeble organic acids: A convenient rapid route to elusive new superbases. *J. Am. Chem. Soc.* 95(3): 982–3, 1973.
10. Bordwell, F. G., *Pure Appl. Chem.* 49: 963–8, 1977; Bordwell, F. G. and Algrim, D. J., *J. Am. Chem. Soc.* 110: 2964, 1988.
11. Greenwood, N. N. and Earnshaw, A., *Chemistry of the Elements, 2nd ed.,* pp. 102–6. Oxford, U.K.: Butterworth-Heinemann, 1997.
12. Eberhardt, G. G. and Butte, W. A., A catalytic telomerisation reaction of ethylene with aromatic hydrocarbons. *J. Org. Chem.* 29(10): 2928–32, 1964.
13. Kamienski, C. W., McDonald, D. P., Stark, M. W., and Papcun, J. R., Lithium and lithium compounds. In *Kirk-Othmer Encyclopedia of Chemical Technology, 5th ed.,* Vol. 15. New York: John Wiley & Sons, 2006.
14. Hoye, T. R., Eklov, B. M., and Voloshin, M., No-D NMR spectroscopy as a convenient method for titering organolithium (RLi), RMgX, and LDA solutions. *Org. Lett.* 6(15): 2567–70, 2004.
15. Lochmann, L., Pospíil, J., and Lím, D., On the interaction of organolithium compounds with sodium and potassium alkoxides: A new method for the synthesis of organosodium and organopotassium compounds. *Tetr. Lett.* 7(2): 257–62, 1966.
16. Schlosser, M. and Hartmann, J., Transmetalation and double metal exchange: A convenient route to organolithium compounds of the benzyl and allyl type. *Angew. Chem. Intern. Ed.* 12(6): 508–9, 1973.
17. Baston, E., Maggi, R., Friedrich, K., and Schlosser, M., Dimetalation: The acidity of monometalated arenes towards superbasic reagents. *Eur. J. Org. Chem.* 2001(21): 3985–3989, 2001.
18. Gilman, H., Beel, J. A., Brannen, C. G., Bullock, M. W., Dunn, G. E., and Miller, L. S., The preparation of *n*-Butyllithium. *J. Am. Chem. Soc.* 71(4): 1499–1500, 1949.
19. Gilman, H. and Cartledge, F. K., The analysis of organoli66thium compounds. *J. Organomet. Chem.* 2(6): 447–54, 1964.

14 Carbon Disulfide

SUMMARY

- Acetylene reacts with liquid sulfur at 360°C producing CS_2 at the rate of ~0.8 mol/h in a 250-mL flask, at a yield of 84% with respect to acetylene, and 76% with respect to sulfur.
- HNMR analysis shows ~3.5 wt% thiophene, C_4H_4S (bp 84°C).
- Other organics and hydrogen compounds are below ~0.5%.

APPLICATIONS [1]

- Reaction with primary amines yields isothiocyanates, which can be reacted further to yield symmetric-substituted thioureas (thiocarbamides). Ammonia or urea yields the thiocyanate, which above 150°C isomerizes to thiourea.
- Treatment of CS_2 with alcoholic NaOH or KOH yields xanthate salts, RO(S)SNa, which react with alkyl chlorides to yield the corresponding xanthate ester.
- Xanthate esters undergo pyrolitic syn-elimination in the Chugaev reaction, yielding alkenes, avoiding skeletal rearrangement or double-bond migration.
- Acidification of cellulose xanthate dissolved in aqueous alkali produces rayon fibers [2].
- CS_2 reacts with aqueous alkali to yield trithiocarbonate salts (Na_2CS_3) [2].
- Reaction with azides produces azidodithiocarbonates $MSCSN_3$, and acidification yields $HSCSN_3$, while oxidation produces azidocarbondisulfide [3].
- Reaction with water at raised temperature yields carbonyl sulfide (COS), and reduction with zinc yields the subsulfide (C_3S_2) [4].
- Chlorination produces thiophosgene and carbon tetrachloride [5].
- Hydrogen-free solvent in spectroscopy.
- Very high solvating power for sulfur, phosphorus, etc. [2].

14.1 INTRODUCTION

Carbon disulfide is a useful solvent and reagent, which is becoming increasingly difficult and expensive to purchase because of a decline in its traditional use. This is compounded by its high level of toxicity and one of the lowest autoignition temperatures for a stable substance. As a result, it is a sea-freight item and most production is local [2]. Preparation methods described in the literature are few and applicable only at the industrial scale. An older method consists of the passage of sulfur vapor through charcoal at 1000°C, while a more modern version is the reaction of sulfur with methane at 650°C catalyzed by alumina. We present a new laboratory method based on the reaction of acetylene with liquid sulfur in a two-neck flask at an oven temperature of 360°C. The method has an efficiency of about 84%, and the production rate of CS_2 is on the order of 0.8 mol/h.

The uses of carbon disulfide in the laboratory are quite diverse, ranging from applications based on its solvating power (it dissolves up to 63 wt% sulfur and 90 wt% phosphorus [2]) to the Chugaev reaction for forming alkenes from alcohols, which avoids double-bond migration or skeletal rearrangement. Its derivatives including trithiocarbonates, xanthates, dithiocarbamates, and thiocarbimides all find wide application in coordination chemistry, which for CS_2, unlike CO_2, is quite diverse [6].

14.2 DISCUSSION

Lampadius first prepared carbon disulfide in 1796 by heating pyrites with charcoal [7], but provided only a terse description of it as an extremely volatile liquid containing sulfur. In 1812, Marcet and Berzulius worked out the elemental constituents of the "sulphuret of carbon" and discovered its potent solvating properties for nonpolar compounds. This established its use throughout the nineteenth century, especially as a solvent in the cold vulcanization of rubber.

Until the 1950s, CS_2 was prepared by the original method of passing sulfur vapor over specially prepared carbon, heated either in retorts [8], or electrothermally [9,10] to 900°C–1000°C [7–10]. In both cases, this is accompanied by the rapid corrosion of the reaction vessel and loss of yield. However, the high reaction temperature is not a consequence of thermodynamics, because the equilibrium constant for the endothermic reaction between carbon and sulfur

$$C(graphite) + 2S(g) \rightarrow CS_2(g), \tag{14.1}$$

approaches unity at about 460°C. It is, rather, due to kinetics and the low concentration of the S_2 reactive species in sulfur vapor at low temperatures [7].

In 1944, Thacker and Miller [8] described a process for producing carbon disulfide by reacting methane with sulfur vapor catalyzed by activated alumina or silica gel, with a conversion efficiency with respect to methane of 94% at 680°C,

$$CH_4 + 4S(g) \rightarrow CS_2(g) + 2H_2S(g), \tag{14.2}$$

where the sulfur is present in an equilibrium mixture of various sulfur chains and rings S_n depending on the reaction temperature. While the temperature for Reaction 14.2 is still much higher than the theoretical equilibrium temperature of 220°C, it is preferable to the heterogeneous coke process. Although Thacker and Miller mention a lower temperature acetylene process, they described it as disadvantageous based on what they considered to be a low conversion of acetylene to CS_2.

In 1928, Pell and Robinson [11] analyzed the reaction between acetylene and liquid sulfur at 325°C and gaseous sulfur at 500°C and 625°C. In each case, acetylene was passed into a heated flask arranged for distillation. The reaction commenced with a flame and proceeded with the accumulation of a brown oil in the receiver. It was found that 38%, 74%, and 77% of the sulfur reacted at 325°C, 500°C, and 625°C, respectively. An initial distillation of the product to a maximum temperature of 150°C produced a distillate, which completely volatilized below 65°C, and which subsequent fractional distillation through a 4-ft column separated into a carbon disulfide fraction, and thiophene, bp 84°C. The relative proportions of CS_2, thiophene (C_4H_4S), and "thiophten" ($C_6H_4S_2$) were reported relatively constant with temperature, being 77%, 12%, 6%, respectively, for the 325°C and 500°C runs, rising to 83% CS_2 at 650°C.

We present here an improved version of the low-temperature acetylene-sulfur process that proceeds at 360°C, and has been found optimal from the viewpoint of minimum volatilization of unreacted sulfur and maximum acetylene conversion. This temperature is easily achieved using an ordinary laboratory oven, and gives a CS_2 yield of 84% based on acetylene and 76% based on sulfur. Not only is the yield higher than reported in the literature [11], but the much lower fractions of C_4H_4S and $C_6H_4S_2$ make purification simpler.

Formally, the main reaction can be written as:

$$C_2H_2(g) + 5S(l) \rightarrow 2CS_2(g) + H_2S(g), \Delta H_{298} = -13.2 \text{ kJ/mol}, \quad (14.3)$$

which is exothermic despite the positive enthalpy of formation of CS_2, $\Delta_f H_{298} = 89$ kJ/mol. The entropy change on the balance of gaseous products below the bp of sulfur also favors the right-hand side, $\Delta G_{300} = -103$ kJ/mol, $\Delta G_{600} = -180$ kJ/mol. However, the observed flame [11] indicates that most of the interaction takes place in the gas phase. Hence, the reactants are more properly represented by a complex equilibrium of gaseous sulfur chains:

$$nS(l) \rightarrow xS_8(g) + yS_6(g) + zS_2(g), \quad (14.4)$$

and calculation of free energy change is more complicated. Nonetheless, the equilibrium expressed in Reaction 14.4 applies to all three CS_2 production processes; hence, in principle, the lower reaction temperature compared to Reaction 14.1–Reaction 14.2 is explained by the high positive $\Delta_f H$ of acetylene.

The acetylene is sourced from a standard welding-grade cylinder (2% maximum impurity [12]). Because acetylene can explosively decompose into its elements at high pressures, it is stored as a solution in acetone. Because a nearly saturated solution occupies 50% more volume than pure acetone, a solid matrix of ~90% porosity is used as a fill, preventing the formation of gaseous acetylene above the solution as acetylene is withdrawn. The acetone forms a maximum 41–43 wt% fraction of the initial charge, and the acetylene pressure varies in the range 1.2–2 MPa (vapor pressure of solution doubles with a 30°C temperature rise [12]). The acetylene volatilizes some of the acetone as well as moisture in the cylinder, so that about 0.1 g acetone per 1 L gas or 10% w/w was measured at the discharge rate of this experiment. To avoid contamination, this is removed by passage of cylinder gas through two scrubbers, the first containing 250 mL of chilled water and the second concentrated sulfuric acid. Acetone is miscible with water, and its vapor pressure at 0°C and 20°C is 9.4 kPa and 24 kPa, respectively [13]; hence, assuming all the acetone dissolves, by Raoults law, 1 mol of acetylene delivers about 0.05% acetone (by weight) to the H_2SO_4 scrubber. While acetylene reacts with sulfuric acid [12], and the latter soon becomes discolored with decomposition products (mainly carbon), this reaction is fairly slow, and less than 1% acetylene is lost in this fashion.

14.3 EXPERIMENTAL

It is recommended that the apparatus be located inside a fume hood, as a precaution in case of malfunction. First, 270 g of flowers of sulfur (99%) is placed in a 250-mL, three-neck flask equipped with a gas inlet tube (~5 mm diameter) extending to within ~⅓ of the flask bottom. The flask is placed in a top-loading temperature-controlled air oven so that its necks protrude several centimeters above the top, and thermal insulation is placed around the flask body so that its contents are maintained at oven temperature, while protecting the neck joints from convective heat from the oven.

A three-way distillation adaptor with a 360°C thermometer greased at the joints is attached to the middle neck, which is at least 24/29, and insulated with aluminum foil and glass wool up to the height of the thermometer bulb (Figure 14.1). A 30-cm Leibig condenser cooled with ice water from a recirculating pump is connected to the adaptor, and this leads the condensed products through a gas take-off adaptor into a 250-mL receiver flask, also cooled in an ice water bath. The gas take-off is connected by means of a silicone hose, via a safety flask, to a 500-mL scrubber containing 300 mL of 20% NaOH solution, which absorbs the toxic H_2S and CS_2 vapors. Low-pressure acetylene from a cylinder is fed via a manometer and flow meter with control valve into a 500-mL scrubber containing 250 mL of water chilled in an ice bath. Gas from this first water scrubber is passed into a concentrated sulfuric acid scrubber with its head lightly plugged with glass wool to trap acid spray. From there, the acetylene is passed into the reaction flask. Because C_2H_2 reacts poorly with NaOH, a small amount of unreacted acetylene exits the apparatus; hence, the exhaust from the NaOH scrubber should be thoroughly vented.

To avoid formation of an explosive $C_2H_2/CS_2/H_2S/O_2$ mixture inside the apparatus, it is thoroughly flushed with nitrogen, connected through a T-junction in the acetylene inlet port. Although acetylene also may be used for this purpose, subsequent heating without positive acetylene pressure in the apparatus can lead to a premature reaction with unpredictable consequences.

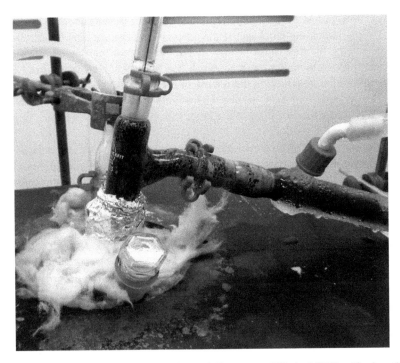

FIGURE 14.1 *A color version of this figure follows page 112.* At 360°C sulfur in a three-neck flask burns in acetylene producing H$_2$S, CS$_2$, and a small amount of C$_4$H$_4$S (thiophene). The two latter condense in the ice water-cooled Liebig condenser and collect in a receiver. About 14% of the sulfur and traces of soot volatilized by the gaseous products also escape the reflux in the vertical section of the adaptor and are carried into the receiver with the CS$_2$.

The oven temperature is now raised to 360°C and held for about 30 min, giving sufficient time for the sulfur to melt and equilibrate, whereupon acetylene is introduced into the flask at the rate of 340 mL/min (0.85 mol/h), which corresponds to about 3 bubbles/sec. Initially, yellow sulfur fumes volatilized by the acetylene appear in the condenser but, after about 30 sec. these change to brown, indicating the start of the reaction. The condenser and adaptors are progressively covered with a black sulfur layer and, after 3–4 min, drops of a black liquid start collecting in the receiver at the rate of about 0.7 drops/sec. This liquid dissolves the previously formed sulfur layer on the adaptor and condenser walls preventing clogging, while the vapor inside the receiver flask remains fairly clear.

The distillation temperature, as indicated by the adaptor thermometer, rapidly rises to 260°C, whereafter it can continue to rise slowly to about 300°C. It should not be allowed to exceed 320°C as reflux at this temperature leads to excessive amounts of unreacted sulfur passing into the receiver. The reflux temperature can be lowered by loosening or lowering the insulation around the distillation adapter. By the same token, the temperature should not be allowed to drop below about 240°C, since too high a reflux ratio leads to sulfur blocking the flask neck. The manometer pressure gives an indication of possible blockage (although this is normally preceded by a rapid drop in distillation temperature). Normal pressure is about 0.2 psi, with a rise to 0.5 psi indicating blockage. If the adaptor temperature is not controlled and blockage occurs, the flask must be cooled and the adaptor thoroughly cleaned with CS_2 prior to recommencing the reaction. It has been observed that a brief loss of flask hermitization, resulting in the release of $S/CS_2/C_2H_2$ vapor at 360°C, does not automatically lead to ignition, despite the low autoignition temperature of CS_2. This is due likely to the modification of the latter's properties by the presence of a large sulfur component.

After about 10–20 min of operation, the contents of the sulfuric acid scrubber become brown due to decomposition of some of the acetylene, while the NaOH solution develops a crimson coloration tending to dark red at the end of the reaction, due to dissolution of unreacted sulfur in the $NaOH/NaCS_3/Na_2S$ solution.

Completion of the reaction is indicated by condensation ceasing and the distillation temperature dropping to about 230°C, while the manometer pressure remains unchanged. A small amount of sulfur is still present in the flask above the gas inlet line at this stage, but gas passage is too shallow to volatilize enough sulfur for the reaction to proceed. The apparatus is allowed to cool prior to dehermitization to prevent CS_2 fumes from igniting on the hot oven surface.

For a reaction time of 96.5 min, with acetylene flow at 336.6 mL/min (a total of 1.35 mol of acetylene at 20°C), 115.5 g of a black oily liquid were gathered in the receiver with the reaction flask losing 150.1 g of sulfur. Redistillation of the liquid product at atmospheric pressure to a maximum distilling flask temperature of 150°C (at which stage no more product passed over) produced 86.2 g of a light yellow liquid, which distilled entirely in the range 47°C–50°C, showing that the percentage of thiophene impurity is small. The CS_2 yield of this fraction based on acetylene is consequently 84%. Altogether, 120 g of sulfur was lost from the system; hence, the yield based on sulfur is 76% of theoretical (24% of sulfur is lost both by conversion to thiophene in side reactions and as elemental sulfur deposited in the NaOH scrubber).

The CS_2 is expected to contain traces of organic sulfur compounds, dissolved sulfur, H_2S, and water. The organosulfides, in particular, are clearly manifested by the strong mercaptan odor of the product, substantially different from ethereal odor of pure CS_2. A GC/MS analysis, however, shows the major nongaseous impurity is thiophene (C_4H_4S), present at the level of a few percent, while the combined content of all other volatiles is below 1%. A more accurate HNMR analysis gives the thiophene content as 3.5 wt%, while all other hydrogen compounds, including water, are constrained below 0.1 mol% on hydrogen content. A ^{13}C NMR analysis confirms the preceding result regarding thiophene and further constrains other organics to below 0.5%.

Thus, for many purposes, further purification is not necessary; however, if desired, the strong mercaptan odor can be removed by shaking the product with $KMnO_4$ solution [14]. The oxidation should not, however, continue for several hours, because traces of mercaptans present are oxidized within 30 min or so upon vigorous shaking with a saturated $KMnO_4$ solution. After that, the reaction continues with CS_2 and C_4H_4S being oxidized at about the same rate of ~5 g/h for equal volumes of reducer and oxidizer in a 500-mL flask. For purification of the present product, the following procedure has been found adequate.

To 100 mL water in a 500-mL conical flask is added 5 g $KMnO_4$, and the flask is shaken well to dissolve as much of the oxidizer as possible. More permanganate will dissolve during the reaction as $KMnO_4$ in solution is reduced to insoluble MnO_2. Then, 100–200 mL of crude CS_2 is now added, and the flask closed with a greased stopper and swirled vigorously on an electric shaker for about 20 min, initially stopping every 30 sec. or so to relieve possible pressure buildup. The contents of the flask are then poured into a separating funnel (leaving behind any MnO_2) and separated from the top layer of oxidation products and aqueous solution.

REFERENCES

1. Schmitt, C. and Murai, T., Carbon disulfide. In *Encyclopedia of Reagents for Organic Synthesis*, edited by Paquette, L. New York: John Wiley & Sons, 2004.
2. Smith, D. E., and Timmerman, R. W., Carbon disulfide. In *Kirk-Othmer Encyclopedia of Chemical Technology, 5th ed.,* Vol. 4. New York: John Wiley & Sons, 2006.
3. Smith, G. B. L. and Wilcoxon, F., Azidodithiocarbonic acid (azidothio-formic acid) and azidocarbondisulfide. *Inorg. Synth.* 1: 81–84, 1939.
4. Schlesinger, G. G., *Inorganic Laboratory Preparations*, p. 653. New York: Chemical Pub. Co., 1962.
5. Mägerlein, H., Meyer, G., and Rupp, H.-D., A new catalytic process for the preparation of thiocarbonyl chloride (thiophosgene). *Synthesis* (1): 26–27, 1974.
6. Greenwood, N.N., and Earnshaw, A., *Chemistry of the Elements, 2nd ed.,* p. 298. Oxford, U.K.: Butterworth-Heinemann, 1997.
7. Stull, D. R., Thermodynamics of carbon disulfide production. *Ind. Eng. Chem.* 41(9): 1968–1973, 1949.
8. Thacker, C. M. and Miller, E., Carbon disulfide production. *Ind. Eng. Chem. Int. Ed.* 36: 182–4, 1944.
9. Taylor, E. R., *Trans. Am. Electrochem. Soc.* 1: 115–7, 1902.
10. Stansfield, A., *The Electric Furnace: Its Construction, Operation and Uses*. New York: McGraw-Hill, 1914.
11. Peel, J. B. and Robinson, P. L., The reaction between acetylene and sulphur at temperatures up to 650°. *J Chem. Soc.* 10: 2068–70, 1928.
12. Gannon, R. E., Manyik, R. M., Dietz, C. M., Sargent, H. B., Thribolet, R. O., and Schaffer, R. P., Acetylene. In *Kirk-Othmer Encyclopedia of Chemical Technology, 5th ed.,* Vol. 1. New York: John Wiley & Sons, 2006.
13. The National Institute of Standards and Technology (NIST) Virtual Library http://webbook.nist.gov/chemistry/form-ser.html.
14. Furniss, B. S., Hannaford, A. J., Smith, P. W. G., and Tatchell, A. R., *Vogel's Textbook of Practical Organic Chemistry, 5th ed.,* p. 412. London: Addison Wesley Longman, 1989.

15 Chlorine

SUMMARY

- Trichloroisocynauric acid is reduced with HCl to yield Cl_2 with low O_2 and H_2O content.
- Quantitative yield makes the amount of chlorine generated easy to control.
- Low content of O_2 enables Cl_2 to be used in thermal and photolytic chlorination.
- Single P_2O_5 desiccation stage is possible due to low H_2O content.

APPLICATIONS

- General-purpose chlorinating agent in synthesis, for example, alkyl/acyl chlorides
- Formation of moisture-sensitive/Lewis acid chlorides, for example, $CrCl_3$, $SnCl_4$, $SiCl_4$, etc. [1]

15.1 INTRODUCTION

Chlorination can often be conducted without resorting to gaseous chlorine, for example, using such reagents as sulfuryl, thionyl, or acetyl chlorides [2]. However, in certain circumstances, these reagents are either insufficiently active or produce by-products that are difficult to separate. Gaseous chlorine dissociated by UV/blue light (continuous absorption band centered at 330 nm, absorbance at 1 atm 0.0375 cm^{-1}, 0.21 cm^{-1}, 2.1 cm^{-1} at 450 nm, 400 nm, and 350 nm, respectively [3,4]), has often been used as a chlorinating agent in the past [5,6], but is becoming increasingly difficult to obtain in cylinder form due to changes in shipping regulations.

Generation in the laboratory generally follows textbook preparation methods based either on the reduction of hydrochloric acid by oxidizing agents, such as $KMnO_4$ and MnO_2, or on the oxidation of hypochlorite. Due to the substantial enthalpy change associated with these reactions, they are rather cumbersome and are difficult to maintain at the constant rate required to produce a steady flow of gas. They also generate a substantial amount of contaminants, such as oxygen, hydrochloric acid, and water vapor. Presented is a simple method based on the co-proportionation of trichloroisocyanuric acid $C_3N_3O_3Cl_3$ (TCCA, a chlorimide) with hydrochloric acid in the following reaction-driven essentially by entropy and, hence, not associated with substantial evolution of heat

$$+ 3HCl \longrightarrow \qquad + 3Cl_2 \tag{15.1}$$

The reaction proceeds at room temperature, without heating, so that the amount of water vapor and HCl volatilized with the chlorine is substantially reduced, making desiccation with a single P_2O_5 stage possible (0.03 mL H_2O/1 L Cl_2, compared to 2–3 mL H_2O/1 L Cl_2 for hypochlorite). At the same time, very little oxygen is generated (<0.3% O_2 compared to 1.6% for hypochlorite). This is important as oxygen strongly inhibits both thermal [7] and photolytic [8] chlorination. The reaction is quantitative, so the addition of a calculated amount of chlorine is easy to implement.

15.2 EXPERIMENTAL

To start, 232 g (1 mol) of TCCA is powdered in an electric grinder and placed in 1-L, two-neck, flat-bottom flask equipped with a 2–3 cm magnetic stirrer. Then, 330 mL of 37% w/vol hydrochloric acid (3 mol HCl), diluted with 330 mL of water, is poured into a pressure-equalized dropping funnel and run into the TCCA at a rate not exceeding 2 drops/sec. The generated chlorine is dried by passage through about 3–4 cm of P_2O_5 in a drying tube and used directly. The generator produces 69 L (2.90 mol) of chlorine gas in 200 min, corresponding to a 97% yield. Heating of the reaction mixture liberates a further 2% chlorine; however, the water vapor and O_2 content of this chlorine is substantially increased.

The slow magnetic stirring has little influence on gas generation in the initial stages of the reaction when the reagents are dry, but helps maintain the rate of chlorine evolution toward the end as well as preventing the formation of unreacted lumps.

REFERENCES

1. Schlesinger, G. G., *Inorganic Laboratory Preparations*, p. 20, 55, 129, 133, 111–121. New York: Chemical Pub. Co., 1962.
2. Furniss, B. S., Hannaford, A. J., Smith, P. W. G., and Tatchell, A. R., *Vogel's Textbook of Practical Organic Chemistry, 5th ed.*, p. 723, 864. London: Addison Wesley Longman, 1989.
3. Seery, D. J. and Britton, D., The continuous absorption spectra of chlorine, bromine, bromine chloride, iodine chloride, and iodine bromide. *J. Phys. Chem.* 68 (8): 2263–66, 1964.
4. Matsumi, Y., Kawasaki, M., Sato, T., Kinugawa, T., and Arikawa, T., Photodissociation of chlorine molecule in the UV region. *Chem. Phys. Lett.* 155(4–5): 486–490, 1989.
5. Vogel, A. I., *Practical Organic Chemistry, 3rd ed.*, p. 538. London: Longman, 1956.

6. Eckert, H. and Forster, B., Triphosgene, a crystalline phosgene substitute. *Angew. Chem., Int. Ed. Engl.* 26(9): 894–5, 1987.

7. Vaughan, W. E. and Rust, F. F., The high temperature chlorination of paraffin hydrocarbons. *J. Org. Chem.* 5(5): 449–71, 1940.

8. Dickinson, R. G. and Leermakers, J. A., The chlorine-sensitized photo-oxidation of tetrachloroethylene in carbon tetrachloride solution. *J. Am. Chem. Soc.* 54(10): 3852–62, 1932.

16 Carbon Tetrachloride

SUMMARY

- Carbon tetrachloride is produced by photolytic chlorination of chloroform at 0.5 mol/h in a 1-L flask with 1-kW mercury vapor lamp illumination.
- Yield is >95% with respect to chloroform.
- Carbon tetrachloride >99% pure; main impurity ~0.1% chloroform at 100% Cl_2 excess.
- Detailed reaction rate analysis is presented and compared to experiment.

APPLICATIONS

- Solvent in spectroscopy with high solvating power and no proton signal in NMR.
- Useful diluent in IR spectrometry. Inert and transparent in useful spectral regions [1].
- Analyte in GC/MS with low bp.
- Solvent for Br_2 in qualitative tests for unsaturation. Low HBr solubility differentiates substitution and addition reactions.
- Solvent in photo-oxidations, for example, trichloroethylene [2], dimethyl carbonate (Chapter 17).
- Thermal chlorinations, for example, of rare earth, transition, and actinide metal oxides [3–5].
- Reagent in the synthesis of CF_4, CBr_4, CI_4, and corresponding mixed halides (freons) by substitution reactions with the corresponding hydrogen halides [6].
- Etching and carbon doping of GaAs substrates [7].
- Laboratory preparation of phosgene by oxidation with H_2SO_4 [8] or H_2O_2 [9].
- C_2Cl_2 generation by reaction with copper in DMF in a one-electron transfer process [10].
- C_2Cl_2 intermediate produced by the reaction between CCl_4 and magnesium, used in the halogenation of cyclopropanes [11].
- Friedel-Crafts acylating agent [12].
- Photolytic addition to olefins [13–14].
- The two-component phosphane–CCl_4 system is used to perform halogenation, dehydration, and P-N bond formation, in the Appel reaction [15].
- Chain transfer agent for moderating the length of polymer chains in vinyl polymerization.

16.1 INTRODUCTION

Carbon tetrachloride is a very useful nonhydrocarbon solvent and reagent, which, similar to CS_2, is becoming progressively more difficult to purchase. This is due to a decline in its traditional use as feedstock in the production of chlorofluorocarbons, the latter being phased out under the Montreal Protocol (1989). Descriptions of laboratory preparations are almost nonexistent because of its low cost and availability in the past. The two main industrial methods of manufacture are exhaustive high-temperature (450°C–650°C) chlorination of hydrocarbons, especially methane [16], and chlorination of CS_2 at 100°C–130°C [17–18]. Both are gas-phase reactions producing a large amount of by-products, and thus they cannot be run as efficient batch processes in the laboratory. A prospective laboratory method is the original photolytic chlorination in distilling chloroform [19] (which is readily available, or can be prepared in the haloform reaction). However, the literature states that chloroform chlorinates with difficulty [19], and very little quantitative information is available that is of relevance to a laboratory preparation. A major concern is the extent of conversion achievable, the excess of chlorine required, the reaction rate in a typical laboratory setup, and percentage impurities in the final product.

We present a laboratory method for producing CCl_4 at a rate of about 1 mol per 100 min in a 1-L flask, UV-illuminated with a commonly available 1-kW high-pressure mercury vapor lamp. The yield is >95% with respect to chloroform, the main impurity being approximately 2.0% residual chloroform at 1.5:1 Cl_2 molar excess, and 0.1% chloroform at 2:1 excess. Further purification by distillation requires a good column due to the proximity of the boiling points of CCl_4 (76°C) and $CHCl_3$ (61°C), corresponding to an enrichment ratio [20] of about 1.6 near the CCl_4 boiling point (bp).

16.2 DISCUSSION

Carbon tetrachloride was most likely first prepared by Michael Faraday in 1821 by the thermal decomposition of hexachloroethane vapor (from photolytic halogenation of ethylene in illuminating gas) on silica at red heat [21]. Because tetrachloroethylene (bp 121°C) is the major product of this reaction, Faraday did not recognize the minor carbon tetrachloride component, but noted that "carbon sesquichloride" commenced boiling at 77°C. In a set of follow-up experiments in 1840, Victor Regnault chlorinated chloroform in sunlight [19], and remarked on a new compound, similar to Faraday's sesquichloride, with a bp of 78°C, a C:Cl ratio of 1:4, and a molecular formula of C_2Cl_8 (due to an error in the theory of gases at that time). He commented on the difficulty of substituting the hydrogen in chloroform, requiring multiple chloroform distillations from a tubulated retort in a chlorine flux until no more HCl is evolved. Subsequently, Kolbe described the formation of carbon tetrachloride and S_2Cl_2 by the chlorination of CS_2 at red heat [22], while in 1860 A. W. Thomson proposed a laboratory version of the same reaction at 100°C using a $SbCl_5$ catalyst [23], which he claimed to be superior to Regnault's method only in the absence of sunlight.

The present preparation is a modernized version of Regnault's method with the multiple distillations replaced by reflux, and a high-pressure mercury vapor lamp serving as the UV source. The main reaction, represented by the simple formula

$$CHCl_3 + Cl_2 \rightarrow CCl_4 + HCl, \tag{16.1}$$

belies a complex multistage process initiated by photolytic dissociation of chlorine into free radicals, followed by free radical propagation by collisions with neutral atoms, and termination by radical recombination [24]. A more detailed quantitative description is presented in Table 16.1, where implicit in each propagation process x is a reverse process x' (rate k'); I is the local intensity; and we have neglected the $CHCl_2$· radical as the rate of its formation is negligible [24].

Calculation of the rate equation from Table 16.1 is simplified by radical concentrations being much smaller than those of neutral molecules. A confirmation of this in the present experiment comes from the small final concentration of hexachloromethane (see Appendix, Figure A.16) despite the large value of k_e, $[C_2Cl_6]/[CCl_4] \sim 0.1\%$, since it can only form in radical–radical collisions. In solving a–f, one is justified therefore in setting the concentrations of free radicals to their steady-state values (which they acquire shortly after initiation). A further simplification results from considering the relative contribution of the recombination processes d, e, f, and l to the chlorine radical concentration. Process f can occur only in three-body collisions where a third body M carries away sufficient energy for the Cl radicals to form a bound state. With the concentration of neutral molecules $[M] \sim 0.04$ mol L^{-1} in the gas phase and ~ 10 mol L^{-1} in the condensed phase, we see this is less likely than CCl_3·–Cl· recombination, Process d, whose rate law is pseudo two-body, and thus is

---

TABLE 16.1
Elementary Processes in the Photolytic Chlorination of Carbon Tetrachloride

	Reaction	Rate Equation	Rate Constants, L mol⁻¹ s⁻¹
Initiation			
a	$Cl_2 + photon \rightarrow 2Cl$·	$R_a = k_a\,I\,[Cl_2]$	
Propagation			
b	$CHCl_3 + Cl$· $\rightarrow CCl_3$· $+ HCl$	$R_b = k_b[Cl·]$ $[CHCl_3]$	$k_b \sim 12$, $k_b' \sim 2\ 10^{-7}$
c	$Cl_2 + CCl_3$· $\rightarrow CCl_4 + Cl$·	$R_c = k_c[Cl_2][CCl_3·]$	$k_c \sim 0.07$, $k_c' \sim 10^{-9}$
Termination			
d	CCl_3· $+ Cl$· $\rightarrow CCl_4$	$R_d = k_d[CCl_3·][Cl·]$	$k_d \sim 870$
e	CCl_3· $+ CCl_3$· $\rightarrow C_2Cl_6$	$R_e = k_e[CCl_3·]^2$	$k_e \sim 870$
f	Cl· $+ Cl$· $+ M \rightarrow Cl_2 + M$	$R_f = k_f[M][Cl·]^2$	$k_f \sim 20\ [M]\ (mol\ L^{-1})$
l	Cl· $+ M_l \rightarrow \frac{1}{2}Cl_2$	$R_l = k_l[Cl·]$	

Source: Cabrera et al., *AIChE J.* 37(10): 1471–84, 1991.

less constrained by kinetic energy phase space. Hence, the free radical concentrations take the following form:

$$[Cl\cdot] \sim \frac{2k_a I[Cl_2]}{k_d[CCl_3\cdot] + k_l} \tag{16.2}$$

and

$$[CCl_3\cdot] \sim [Cl\cdot] \frac{k_b[CHCl_3] + k_c'[CCl_4]}{k_b'[HCl] + k_c[Cl_2]}, \tag{16.3}$$

where we have used $k_d[Cl\cdot] \ll k_c[Cl_2]$ in Equation 16.3. If this were not the case in the present experiment, it would follow from the respective rate laws that the amount of CCl_4 formed in the termination step d is not much less than in the propagation step c. Then using $[CCl_3\cdot] > [Cl\cdot]$, which follows from Equation 16.3 (even in its exact form), this would mean that the total amount of C_2Cl_6 formed is not much less than CCl_4, contrary to observations.

Process l corresponds to the absorption of chlorine radicals on particle aggregates, such as reactor surfaces. In gases, this is negligible since typical reactor dimensions are much larger than the mean free path for radical recombination $l_f \sqrt{([\text{radicals}]/[\text{neutral species}])} \sim 0.01–1$ mm, with $l_f \sim 0.1$ μm being the mean free path under standard conditions.

In the liquid phase on the other hand, collisions are replaced by continuous interactions, and l is the dominant process at low-to-medium radical concentrations [14,25].

Thus, the overall reaction rate is

$$\frac{d[CCl_4]}{dt} \sim [Cl\cdot] \frac{k_b k_c[CHCl_3][Cl_2] - k_b' k_c'[HCl][CCl_4]}{k_b'[HCl] + k_c[Cl_2]}. \tag{16.4}$$

Substituting Equation 16.2 and Equation 16.3, we obtain the *photochlorination rate equation* far from equilibrium:

$$\frac{d[CCl_4]}{dt} \sim [Cl_2]\sqrt{2k_a k_b k_c I[CHCl_3]/k_d} \text{ gas phase,} \tag{16.5}$$

$$\frac{d[CCl_4]}{dt} \sim k_a k_b I[Cl_2][CHCl_3]/kl \text{ liquid phase.} \tag{16.6}$$

One recognizes the numerator in Equation 16.4 as the standard expression for the equilibrium constant of Reaction 16.1 in terms of the forward and reverse reaction rates. Because equilibrium lies far toward the right ($[CHCl_3]/[CCl_4] < 0.1\%$ in final product), reverse processes can initially be neglected, so in the gas phase the reaction rate is proportional to the square root of the intensity and chloroform concentration, while in the liquid phase there is direct proportionality. These kinds of relationships have been observed by Vaughan and Rust [25]. Because of the different dependencies on chloroform concentration in the two cases, we have two different types of dependence of concentration on time far from equilibrium, as demonstrated by the following *rate equation solutions*:

$$[CHCl_3](t) = C_0 e^{-\alpha t} \quad \text{liquid phase,} \tag{16.7}$$

$$[CHCl_3](t) = \left(\sqrt{C_0} - \alpha t/2\right)^2 \quad \text{gas phase,} \tag{16.8}$$

where C_0 is the initial chloroform concentration, and α is the proportionality constant in both cases. These equations are used here to distinguish the two cases experimentally.

16.2.1 THE PHOTOLYTIC EFFICIENCY

We can relate the photolytic rate constant k_a to the primary quantum yield $\Phi(\lambda)$ by the following considerations. The power absorbed per volume of fluid is related to the local intensity I by

$$\frac{dP}{dV} = \frac{I}{L_0} \equiv \alpha[Cl_2]I , \tag{16.9}$$

where L_0 is the absorption length, and α is the extinction coefficient. The absorbed energy dE, corresponding to the number of broken Cl-Cl bond in volume dV, is

$$dE = \frac{\Delta H}{2\Phi(\lambda)}[Cl\cdot]dV , \tag{16.10}$$

where ΔH is the bond energy of the Cl-Cl, bond and $\Phi(\lambda)$ is the fraction of energy absorbed at wavelength λ, which goes toward creating free chlorine radicals. Differentiating Equation 16.10 with respect to time, combining with Equation 16.9, and comparing to the rate equation for Process a in Table 16.1, we find

$$k_a = \frac{\alpha\Phi(\lambda)}{\Delta H} . \tag{16.11}$$

The photolytic rate constant thus is proportional to the primary quantum yield, and this is used below to derive the photolytic efficiency.

A Cl-Cl bond energy of 245 kJ/mol gives, using Planck's law, $\lambda = 488$ nm as the threshold photon wavelength for dissociation, which agrees with the continuous absorption spectrum of chlorine gas commencing at about 500 nm. The spectrum peaks at 335 nm, and Figure 16.1 shows that, under standard conditions, the chlorine absorption length drops from over 30 cm at 480 nm, to less than 1 cm at 360 nm.

The ratio of the absorption length to the characteristic length of the reaction vessel is a critical parameter. A ratio much larger than 1 corresponds to low overall absorption and, hence, low photolytic efficiency, while a ratio much less than 1 leads to most of the reactor volume being devoid of free radicals. While the latter

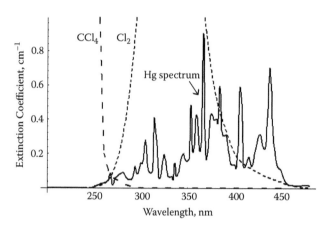

FIGURE 16.1   Wavelength dependence of the inverse absorbance length for Cl_2 and CCl_4 under standard conditions. Superimposed is the intensity spectrum of a typical high-pressure Hg vapor lamp (Heraeus OH 1000) in relative units.

case leads to a corresponding increase in free radical concentration in the active volume, this is not proportional to the decrease in active volume itself because of increased free radical recombination. This is reflected in the square root dependence in Equation 16.5. There results an optimal wavelength maximizing average free radical concentration.

Superimposed in Figure 16.1 is the radiation spectrum of a typical high-pressure mercury vapor lamp in the range 200–450 nm (in which the lamp radiates 70% of its output). The overall reaction rate, which is proportional to average value of $[CCl_4]$, is obtained by integrating Equation 16.5 over this radiation spectrum and reactor volume. Absorbing all factors independent of space and wavelength into a proportionality constant, we get the following relation:

$$\frac{d[CCl_4]}{dt} \propto \frac{1}{\int_{\text{Reactor volume}} dV} \int_{\text{Reactor Volume}} dV \sqrt{\int_{\text{Hg Spectrum}} d\lambda \frac{\Phi(\lambda)I(\lambda)}{L_0(\lambda)} e^{-L(V)/L_0(\lambda)}}, \quad (16.12)$$

where $L(V)$ is the light path length through the reactor as a function of position, and $I(\lambda)$ is the intensity spectrum of the lamp. The reduction in reaction rate at short absorption lengths is evidenced by the peaked nature of the integrand with respect to L_0. The photolytic efficiency is expressed as the ratio of the reaction rate for the spectrum of the mercury lamp, to the reaction rate for a monochromatic light source of the same total power at the optimal wavelength. Using Equation 16.12, assuming $\Phi(\lambda)$ to be slowly varying, and using the spectrum of Figure 16.1, the photolytic efficiency of the present experiment is about 55% for characteristic reactor sizes of $L =$ 5–20 cm. The integral in Equation 16.12, therefore, evaluates to $0.55\sqrt{I\Phi/L}$.

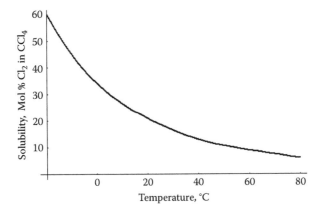

FIGURE 16.2   Variation of the molar solubility of gaseous chlorine in carbon tetrachloride as a function of temperature.

16.2.2   COMPARISON WITH EXPERIMENT

Figure 16.2 shows that chlorine has about 5 mol% solubility in boiling chloroform/ carbon tetrachloride [26], corresponding to a molar concentration in the liquid phase of 0.5 mol/L . This is substantially larger than the gas-phase chlorine concentration of ~0.03 mol/L at 65°C. However, very little useful radiation reaches the liquid phase as the path length through the gas phase of the reactor is larger than the absorbance length for most of the useful spectrum. Hence, the contribution of the liquid phase reaction in the present experiment is minor.

Figure 16.3 shows the variation of percentage chloroform remaining in the product as a function of the mole ratio of introduced chlorine to initial chloroform (the chlorine is introduced at a fixed rate) in the present experiment. The fairly constant reaction rate over much of the curve prior to equilibrium (up to a mole ratio of about 1.3) fits better a square root dependence on chloroform concentration, Equation 16.8, than the exponential decay of Equation 16.7. The form of the rate equation, thus, corresponds better with a gas phase reaction than with a liquid phase.

Chloroform concentration decreases from 2.0% to 0.1% when the mole ratio of added chlorine increases from 1.5 mol to 2.0 mol. On the other hand, the hexachloromethane content formed in Process e increases over this period. Because distillation is required to separate carbon tetrachloride from hexachloromethane, in certain applications it is advantageous to stop the chlorination at an added chlorine mole ratio of 1.5.

16.3   EXPERIMENTAL

Begin by putting 239 g (2 mol) of chloroform in a three-neck, 1-L borosilicate glass flask (UV transmission for 2-mm-thick borosilicate glass drops to 50% at 310 nm, and so has little effect on efficiency) placed on a heating mantle, and equipped with a chlorine inlet tube reaching about 1 cm from the bottom, and a reflux condenser

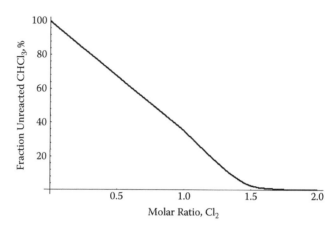

FIGURE 16.3 Observed dependence of the amount of unreacted chloroform in the $CHCl_3$–CCl_4 mixture as a function of the amount of gaseous chlorine passed through the reactor. The latter is expressed as a ratio with respect to the stoichiometric amount required for complete conversion of chloroform.

fed with ice water from a recirculating pump. The outlet of the reflux condenser is passed through a $CaCl_2$ protection tube and over a 10% NaOH solution containing at least 120 g (3 mol) NaOH, prior to being vented.

A 1-kW, high-pressure, Hg vapor lamp (Heraeus OH 1000, cylindrical 50 mm long, 24.5-mm diameter, 140 V, 8.2 A) is placed horizontally 20 cm from the flask and roughly at the same height as the top of the flask. A lower-wattage mercury lamp does not introduce a substantial penalty since, as seen in the discussion, the reaction rate is proportional only to the square root of the radiant power for this gas-phase reaction. Smaller lamps can be compensated for by shortening the distance to the reactor flask (intensity varies as the inverse square of the lamp-to-flask separation up to separations comparable to lamp size, and then flattens out). Airflow from a small (~20 W) fan is directed between the lamp and the flask to cool the former and avoid formation of hotspots on the latter. Glass wool insulation is tied around the greased quickfit joints close to the lamp to protect from UV and heat, which can result in the joint seizing.

A chlorine gas generator of the type described in Chapter 15 is connected to the gas inlet through a concentrated H_2SO_4 scrubber, containing about 100 mL H_2SO_4, and carrying a glass wool plug in the scrubber head to absorb acid spray. If a 1-L flask is used in the generator (this holds ~232 g, or 1 mol TCCA), it needs to be recharged with approximately 150 g of fresh TCCA toward the end of the procedure (after about 2.5 mol Cl_2 has been added). The entire setup is shown in Figure 16.4.

The apparatus is flushed with nitrogen through the gas inlet at the top of the chlorine generator, the heating mantle is turned on, and the chloroform brought to a gentle reflux (about 2 drops/sec., corresponding to about 10% duty cycle on the 380 W heating mantle shown). The UV lamp is started, and chlorine admitted

FIGURE 16.4 *A color version of this figure follows page 112.* Experimental setup for photolytic chlorination of chloroform. The chlorine generator, described in Chapter 15, is on the left in the photo. The 1-kW, high-pressure, mercury vapor lamp is on the right.

into the reaction vessel at a rate corresponding to about 0.5 drops/sec. HCl in the generator. For chlorine fed at this rate, there should be a noticeable difference in the depth of color between the flask and the top of the condenser, demonstrating a significant amount of chlorine absorption. Chlorine can be tested at the condenser vent by briefly passing the exhaust gases through a saturated NaCl solution, in which chlorine is insoluble, while HCl dissolves freely. This produces a white jet of precipitated NaCl (as NaCl is insoluble in concentrated HCl solution). The rate of chlorine bubbling observed in this way should be about one third of the rate in the H_2SO_4 scrubber.

As the chlorination progresses, the rate of reflux diminishes due to an increasing concentration of CCl_4 and a corresponding increase in the bp of the mixture. Heating is increased to compensate for this, with about 15% duty cycle required at the end of the procedure. After 4 mol of Cl_2 has been introduced (4–5 h), the rate of chlorine evolution (measured as before) should significantly increase, and the product should acquire a light yellow coloration that does not disappear after the UV lamp has been turned off for about 1 min.

The density of the product at 20°C is 1.587 g/cm^3 (sp gr CCl_4 = 1.587, sp gr $CHCl_3$ = 1.483). GC/MS and NMR analyses (Appendix, Figure A.16 and Figure A.17), show that the product contains ~0.1% $CHCl_3$ and ~0.1% C_2Cl_6, the remainder being CCl_4.

REFERENCES

1. Streitwieser, A. and Heathcock, C. H., *Introduction to Organic Chemistry*, pp. 338. New York: Macmillan, 1976.
2. Dickinson, R. G. and Leermakers, J. A., The chlorine-sensitized photo-oxidation of tetrachloroethylene in carbon tetrachloride solution. *J. Am. Chem. Soc.* 54(10): 3852–62, 1932.
3. Reed, J. B., Hopkins, B. S., Audrieth, L. F., Selwood, P. W., Ward, R., and Dejong, J. J., Anhydrous rare-earth chlorides. *Inorg. Synth.* 1: 28–33, 1939.
4. Hein, F. and Herzog, S., Chromium, molybdenum, tungsten, uranium. In *Handbook of Preparative Inorganic Chemistry, 2nd ed.*, p. 965, edited by Brauer, G. New York–London: Academic Press, 1963.
5. Li, Y., Qian, Y., Liao, H., Ding, Y., Yang, L., Xu, C., Li, F., and Zhou, G., A reduction-pyrolysis-catalysis synthesis of diamond. *Science* 281(5374): 246–7, 1998.
6. Soroos, H. and Hinkamp, J. B., The redistribution reaction. XI. Application to the preparation of carbon tetraiodide and related halides. *J. Am. Chem. Soc.* 67(10): 1642–3, 1945.
7. Li, L., Gan, S., Han, B.-K., Qi, H., and Hicks, R. F., The reaction of carbon tetrachloride with gallium arsenide. *Appl. Phys. Lett.* 72(8): 951–3,1998.
8. Gymer, G. E. and Narayanaswami, S. Functions containing a carbonyl group and at least one halogen, p. 412–3. Vol. 6 of *Comprehensive Organic Functional Group Transformations*, edited by Katritzky, A. R., Moody, C. J., Meth-Cohn, O., and Rees, C. W. New York: Pergamon, 1995.
9. Tatarova, L. A., Trofimova, K. S., Gorban, A. V., and Khaliullin, A. K., Reaction of carbon tetrachloride with hydrogen peroxide. *Russ. J. Org. Chem.* 40(10): 1403–6, 2004.
10. Egorov, A. M., Matyukhova, S. A., and Anisimov, A. V., Kinetics and mechanism of the reaction of carbon tetrachloride with copper in dimethylacetamide. *Kinet. Catal.* 44(4): 471–5, 2003.
11. Lin, H., Yang, M., Huang, P., and Cao, W., A facile procedure for the generation of dichlorocarbene from the reaction of carbon tetrachloride and magnesium using ultrasonic irradiation. *Molecules* 8(8): 608–13, 2003.
12. Furniss, B. S., Hannaford, A. J., Smith, P. W. G., and Tatchell, A. R., *Vogel's Textbook of Practical Organic Chemistry, 5th ed.*, pp. 1009–10. London: Addison Wesley Longman, 1989.
13. Harfoush, A., Kinetic study of the addition reaction of carbon tetrachloride to 1-Hexene initiated by chromium hexacarbonyl in the presence of UV light. *Ind. J. Chem. Sect. A* 44(7): 1359–64, 2005.
14. Okamoto, H., Adachi, S., Takada, K., and Iwai, T., Mechanism of radiation-induced addition reaction of carbon tetrachloride onto liquid 1,2-polybutadiene accompanied by cyclization. *J. Polymer Sci.* 17(5): 1279–85, 1979.
15. Appel, R., Tertiary phosphane/tetrachloromethane, a versatile reagent for chlorination, dehydration, and PN linkage. *Angew. Chem. Int. Ed. Engl.* 14(12): 801–11, 1975.
16. Holbrook, M. T., Carbon tetrachloride. In *Kirk-Othmer Encyclopedia of Chemical Technology, 5th ed.*, Vol. 5. New York: John Wiley & Sons, 2006.
17. Mägerlein, H., Meyer, G., and Rupp, H.-D., A new catalytic process for the preparation of thiocarbonyl chloride (thiophosgene). *Synthesis* (1): 26–27, 1974.
18. James, J. W., The action of chlorine on organic thiocyanates. *J. Chem. Soc. Trans.* 51: 268–74, 1887.
19. Regnault, V., Über die Einwirkung des Chlors auf die Chlorwasserstoffäther des Alkohols und Holzgeistes und Über Mehrere Punkte der Aether-Theorie. *Eur. J. Org. Chem. (Ann. Chem. Pharm.)* 33(3): 310–34, 1840.
20. Hill, J. B. and Ferris, S. W. Laboratory fractionating columns. *Ind. Eng. Chem.* 19: 379–82, 1927.
21. Faraday, M., *Ann. Chim. Phys.* 18: 55, 1821.

22. Kolbe, H., Notiz Über einige Gepaarte Verbindungen der Chlorkohlenstoffe. *Eur. J. Org. Chem. (Ann. Chem.)* 49(3): 339–41, 1844; Beiträge zur Kenntniss der Gepaarten Verbindungen. *Eur. J. Org. Chem. (Ann. Chem.)* 54(2): 145–88, 1845.

23. Hofmann, A. W., Über das Verhalten des Schwefelkohlenstoffs zum Antimonchlorid. *Eur. J. Org. Chem. (Ann. Chem.)* 115(3): 264–7, 1860.

24. Cabrera, M. I., Alfano, O. M., and Cassano, A. E., Nonisothermal photochlorination of methyl chloride in the liquid phase, *AIChE J.* 37(10): 1471–84, 1991.

25. Vaughan, W. E. and Rust, F. F., The high-temperature chlorination of paraffin hydrocarbons. *J. Org. Chem.* 5(5): 449–71, 1940.

26. Schmittinger, P., *Chlorine: Principles and Industrial Practice*, p. 9. Weinheim: Wiley-VCH Verlag, 2000.

17 *Bis*-Trichloromethyl Carbonate (Triphosgene)

SUMMARY

- Triphosgene 149 g (½ mol) is generated in a 250 mL reactor by a 6-h photolytic chlorination of dimethyl carbonate (DMC), representing a substantial time saving.
- Yield with respect to DMC is 97%, with >97.5% exhaustively chlorinated final product (HNMR).

APPLICATIONS [1]

Substitute for phosgene
- Chlorination of carboxylic acids and metal oxides.
- Chloroformylation of amines and alcohols, proceeding to isocyanates for primary amines.
- Carbonylation of alcohols to carbonate esters and α-amino acids to *N*-carboxyanhydrides.
- Dehydration of amides to nitriles and organic acids to acid anhydrides.

Specific triphosgene reactivity
- Synthesis of asymmetric ureas and carbonates.

17.1   INTRODUCTION

Phosgene is an unusually versatile reagent realizing a wide range of transformations. Despite this, it is very much underused in the laboratory due to its low bp (7.6°C), high latent toxicity, and inadequate odor warning. Two main substitutes have been proposed in the literature to circumvent these difficulties: trichloromethyl chloroformate, CCl_3CO_2Cl [2], which is formally a dimer of phosgene, and *bis*-trichloromethyl carbonate, $CCl_3CO_3CCl_3$ [3], which is formally a trimer. Both decompose thermally and catalytically to phosgene by a well-known mechanism [4], and this can be used to generate phosgene in situ. On the other hand, the reaction can proceed directly, with triphosgene in particular showing some specific reactivity [1]. Due to a lower vapor pressure, triphosgene is generally the preferred substitute.

One of the drawbacks of triphosgene is its high cost, while the literature preparation is laborious, requiring 28 h of photolytic chlorination (yielding ½ mol [3]). The present experiment demonstrates a substantial reduction in chlorination time to 6 h, producing 149 g (½ mol) triphosgene in a photolytic chlorination of 46.5 g dimethyl carbonate in a 250-mL reactor. This represents a 97% yield, with >97.5% fully chlorinated final product (NMR). The setup uses a single 1-kW mercury vapor lamp, and does not require external stirring or special apparatus [5]. Raising the temperature at the end of the preparation to prevent crystallization is also not required. If desired, dimethyl carbonate can be formed by chloroformylation of methanol, producing 3 mol of DMC for every mole of triphosgene consumed [6], thus providing a synthetic route to triphosgene from methanol and chlorine as starting materials.

17.2 DISCUSSION

Triphosgene was first prepared by Councler in 1880 [7], by photolytic chlorination of dimethyl carbonate to solidification, followed by drying in a vacuum over sulfuric acid (the dimethyl carbonate was first prepared by a lengthy reflux of methyl chloroformate over lead oxide). Councler noted its substantial volatility, and remarked that below its bp (203°C) there was no evidence for decomposition to CCl_4:

$$(CCl3)_2CO_3 \rightarrow CCl_4 + CO_2 + COCl_2, \qquad (17.1)$$

which he expected by analogy with the ethyl compound. This decomposition, which is favored by a six-membered transition state, has recently found IR support [6].

In a more recent preparatory method [3,6], dimethyl carbonate was photochlorinated using two 300 W mercury lamps at 10°C–20°C in CCl_4 solution, in which chlorine is very soluble (20 mol% at 20°C; Figure 16.2), followed by separation of the solvent in vacuum. Liquid chlorination is undoubtedly more efficient and effective than Councler's method, resulting in a 97% yield [3]. Further cooling to 5°C–10°C [5] has been suggested as a way to decrease reaction time by >30% due to the increased chlorine solubility at lower temperatures [6]. The Cl_2 concentration of the saturated solution from Figure 16.2 is about 3 mol/L, which from the data in Figure 16.1 corresponds to an absorption length of less than 20 μm below 360 nm. The reaction in such a solution occurs in a small surface layer, and recombination by Process f of Table 16.1 is expected to be important, so that the reaction rate varies only as the square root of the intensity. Moreover, saturated solutions are not achieved during photochlorination either in the literature [3], or in the present experiment, as the reaction at these concentrations is extremely rapid, and the steady-state chlorine concentration is well below saturation. Thus, a higher solution temperature of 20°C–30°C is maintained in this experiment without affecting the yield.

An investigation of the electrophilic attack of methanol (0.3 M in $CDCl_3$) by phosgene, diphosgene, and triphosgene (0.01 M), demonstrates a 100-fold decrease in the order of reactivity of the three compounds, evidenced by the following change in the pseudo-first-order rate constants: 1.7×10^{-2}, 9.1×10^{-4}, and 1.0×10^{-4} s^{-1}, respectively [4]. The major reaction products from triphosgene electrophilic attack are trichloromethyl carbonate and methyl chloroformate formed in a 1:1 ratio, suggesting the following tetrahedral intermediate:

$$Cl_3COCOCCl_3 + CH_3OH \longrightarrow Cl_3CO\overset{\displaystyle OCH_3}{\underset{\displaystyle OH}{\overset{|}{\underset{|}{C}}}}OCCl_3 \longrightarrow Cl_3COCOCH_3 + COCl_2 + HCl \quad (17.2)$$

with the methyl chloroformate resulting from a subsequent electrophilic attack of methanol by phosgene. Reaction with diphosgene proceeds via a similar tetrahedral intermediate, but the presence of two good leaving groups, Cl^- and Cl_3CO^-, leads to a different final product distribution, with ~8:3 preponderance of trichloromethyl carbonate.

An important feature of these reactions is that although phosgene is a product of the chloroformylation by triphosgene, it is several orders more reactive than triphosgene and, hence, phosgene does not accumulate in the reaction products even in the absence of a nucleophilic substrate.

Triphosgene can, however, be entirely converted to phosgene by catalytic and thermal decomposition. This is shown by Pasquato et al. [4]. A catalytic amount of a weak nucleophile, such as Cl^-, in solution (0.3 mol% tetrabutylammonium chloride) decomposes triphosgene through the formation of a tetrahedral intermediate similar to Reaction 17.2, followed by loss of the trichloromethoxy ion, yielding phosgene and regenerating the chloride ion. The diphosgene product decomposes faster than triphosgene and, therefore, forms only a low-concentration intermediate. Phosgene is the final product with a rate constant of ~1.2×10^{-4} s^{-1}. Triphosgene also can be rapidly converted to phosgene by mixing with initiators, such as activated charcoal, or Lewis acids, such as $AlCl_3$, and heating above its melting point [8].

While this is an effective means of generating phosgene in the laboratory (replacing the more cumbersome CCl_4/H_2SO_4 reaction [9]), it can lead to unexpected results. Thus, the author has found that thin deposits of triphosgene left on glassware after the evaporation of solvents are not susceptible to hydrolysis by water (a nucleophile) even after several days' exposure, in accord with Cotarca et al. [10], where no significant acidity was observed to be imparted to water by triphosgene. Washing the deposit in acetone, however, immediately generates a substantial amount of phosgene produced by a reaction with water solved in the acetone, the latter thus acting as a cosolvent for water/triphosgene [10].

The thermal stability of triphosgene is subject to some controversy [10]. Decomposition generally proceeds via a four-membered intermediate to phosgene and diphosgene [11].

$$Cl_3COCOCCl_3 \longrightarrow Cl_3CO\overset{\displaystyle O\text{-----}CCl_2}{\underset{\displaystyle O}{\overset{|}{\underset{\|}{C}}}\text{-----}Cl} \longrightarrow Cl_3COCCl + COCl_2 \quad (17.3)$$

A calorimetric investigation has shown a decomposition peak at 160°C with a final temperature of 179°C [6], and a start of decomposition at 131°C. However, the decomposition rate at this temperature is exceedingly low, and in the present experiment triphosgene maintained at 135°C under reduced pressure for several hours to thoroughly volatilize the CCl_4 solvent showed no signs of decomposition. On the other hand, the author has found that triphosgene in CCl_4 solution decomposes instantaneously and completely to phosgene in the injector port of a GC/MS at 208°C, with no peaks other than phosgene observed in a chemical ionization mass spectrum.

17.3   EXPERIMENTAL

A tall 250–300-mL wash bottle fitted with a two- or three-neck adapter is placed inside a wide glass funnel suspended by a ring over a thermally insulated container filled with an ice water mixture. The bottle is fitted through the adapter with a long gas inlet tube reaching within ~1 mm of the bottom, so as to minimize bubble size and maximize mixing. Alternatively, the gas bubbler can be fitted with a coarse glass frit. The other neck of the adaptor is fitted with a double surface reflux condenser cooled with ice water circulated from the container below by means of a 2 W recirculating pump. A tube from the condenser outlet feeds water down into the funnel, thus cooling the wash bottle, and from there the water drips back into the container. This arrangement is shown in Figure 17.1. The water–ice surface is thermally insulated by a layer of glass wool from lamp radiation. The outlet of the condenser is protected from moisture condensation by a $CaCl_2$ tube, and the emerging gas is passed over 1 L of 20% NaOH solution and vented.

A 1-kW, high-pressure, Hg vapor lamp (Heraeus OH 1000, as described in Chapter 16) is placed horizontally 10 cm from the bottle at a height corresponding to about the midpoint. Quickfit joints close to the lamp are wrapped with thermal insulation to prevent heat and UV seizing the joint. Because the reaction is liquid phase, lamp intensity is expected to exert a proportional effect on the reaction rate, in contrast to the previous preparation (see Chapter 16, Equation 16.6).

A chlorine generator is assembled and charged with 270 g TCCA and 700 mL concentrated HCl–water mixture, which is sufficient to produce 3.5 mol Cl_2. The chlorine is dried by passage through a concentrated H_2SO_4 scrubber whose head is plugged with glass wool to trap H_2SO_4 spray.

Then, 46.5 g (0.52 mol) dimethyl carbonate is placed inside the wash bottle followed by 200 mL of dry CCl_4 (similar to toluene, CCl_4 can be dried by distillation, discarding the first 10%). Next, dry nitrogen gas is introduced through the dropping funnel of the gas generator to flush the apparatus of oxygen (which inhibits chlorination; see Chapter 16) and moisture, and the mercury lamp is turned on prior to chlorine entering the reactor to prevent the sudden onset of a vigorous reaction once irradiation commences due a substantial quantity of dissolved chlorine. The HCl drip rate in the generator is adjusted to about 1 drop/sec. and the outlet gases are tested for the absence of chlorine by momentary passage through a concentrated NaCl solution, as described in the CCl_4 preparation.

As the chlorination proceeds, the product evolves as a mixture of various ratios of the eight intermediate chlorinated compounds, which with increasing chlorine content of the methyl radical become progressively more difficult to chlorinate [6]. As the chlorine flow rate remains constant, excess chlorine progressively dissolves in the CCl_4 and the liquid acquires a yellow coloration. The increasing chlorine

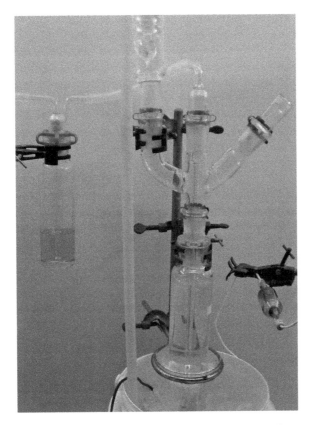

FIGURE 17.1 *A color version of this figure follows page 112.* Setup for photolytical chlorination of dimethyl carbonate in carbon tetrachloride. The free radical reaction is initiated by UV light from a 1-kW, Hg, high-pressure mercury vapor lamp. The bottle reactor is water cooled and the UV lamp is forced-air cooled. This arrangement increases the volume of liquid receiving useful radiation, as the absorption length for UV in the solution is very short.

concentration serves to offset the decrease in the rate constant so that chlorination proceeds at an effectively constant rate throughout the procedure, and no excess chlorine should be evident in the outlet gases until the very end.

The reaction is complete when the yellow chlorine coloration of the reagents persists on exposure to UV light for several minutes with no further chlorine added from the generator. The flask should show a weight gain of 106.5 g. The contents of the reagent bottle are transferred to a 500-mL flask and the CCl_4, evaporated to a final bath temperature of 140°C. After emptying the receiver and allowing the product temperature to drop below 60°C to avoid sudden excessive boiling, the distillation is continued in an aspirator vacuum-dried with P_2O_5, to an ultimate temperature of 135°C held for several hours. A small amount of product is lost due to co-volatilization with the CCl_4, but the final yield is 149 g, or 97% with respect to dimethyl carbonate. Quantitative NMR shows less than 2.5 mol% hydrogen with respect to triphosgene.

As mentioned in the discussion, triphosgene can easily and inadvertently be converted to phosgene, moreover it acts as a cell poison in its own right. Thus, nitrile gloves ordinarily provide adequate protection; however, if the gloves contact traces of triphosgene, which can form by sublimation on quickfit joints, and thereafter inadvertently touch the skin, large red (painless) patches develop after several hours, and take a few days to disappear. Traces of triphosgene on glassware cannot be removed by lengthy soaking in water, and must be washed off with acetone, which immediately produces an easily detectable level of phosgene. Hence, all washing must take place in the fume hood. Bottles containing the solid also develop a substantial phosgene content (without substantial pressure) and must be opened inside the fume hood.

Phosgene test strips (paper soaked in 5% diphenylamine, 5% p-dimethylaminobenzaldehyde alcoholic solution [9]), are a useful safety measure against long-term exposure to low-level phosgene concentrations. For concentrations above the odor threshold (about 0.4 ppm), the olfactory nerves provide a much faster warning, and, hence, it is important to know what to look out for. At low concentrations, phosgene has a pleasant sweet smell resembling air freshener, which is neither suffocating nor resembling moldy hay as is sometimes described. There is no lachrymatory action or soreness of the throat. It is important to take action upon this warning sign immediately, as sensitivity to the odor at this low level rapidly fades.

REFERENCES

1. Roestamadji, J., Mobashery, S, and Banerjee, A., Bis(trichlormethyl) carbonate. In *Encyclopedia of Reagents for Organic Synthesis*, edited by Paquette, L. New York: John Wiley & Sons, 2004.
2. Kurita, K. and Iwakura, Y., Trichloromethyl chloroformate as a phosgene equivalent: 3-Isocyanatopropanoyl chloride. *Org. Synth. Coll.* 6: 715–718, 1988.
3. Eckert, H. Forster, B., Triphosgene, a crystalline phosgene substitute. *Angew. Chem., Int. Ed. Engl.* 26(9): 894–5, 1987.
4. Pasquato, L., Modena, G., Cotarca, L., Delogu, P., and Mantovani, S., Conversion of bis(trichloromethyl) carbonate to phosgene and reactivity of triphosgene, diphosgene, and phosgene with methanol. *J. Org. Chem.* 65(24): 8224–8, 2000.
5. Falb, E., Nudelman, A., and Hassner, A., A convenient synthesis of chiral oxazolidin-2 -ones and thiazolidin-2-ones and an improved preparation of triphosgene. *Synth. Comm.* 23(20): 2839–44, 1993.
6. Cotarca, L., Delogu, P., Nardelli, A., and Sunjic, V., Bis(trichlormethyl) carbonate in organic synthesis. *Synthesis* 553–76, 1996.
7. Councler, C., Kohlensaures methyl. *Ber. Deut. Chem. Ges.* 13(2): 1697–9, 1880.
8. Hales, J. L., Jones, J. I., and Kynaston, W., The infrared absorption spectra of some chloroformates and carbonates. The structure of di- and tri-phosgene. *J. Chem. Soc.* 618–25, 1957.
9. Furniss, B. S., Hannaford, A. J., Smith, P. W. G., and Tatchell, A. R., *Vogel's Textbook of Practical Organic Chemistry, 5th ed.*, p. 457. London: Addison Wesley Longman, 1989.
10. Cotarca, L. and Eckert, H., *Phosgenations—A Handbook*, pp. 21–3. Weinheim: Wiley-VCH Verlag, 2003.
11. Hood, H. P. and Murdock, H. R., Superpalite. *J. Phys. Chem.*, 23(7): 498–512, 1919.

18 Phosphorus Pentachloride

SUMMARY

- Phosphorus pentachloride is prepared by thermal chlorination of technical grade (40–60%) calcium phosphide at 240°C.
- A 22-mm × 20-cm borosilicate glass tube yields 12.3 g PCl_5 in ~30 min, at quantitative yield based on Ca_3P_2 present in the technical phosphide.
- Sublimation yields a product with ~10 wt% PCl_3 due to the $PCl_3 \leftrightarrow PCl_5$ equilibrium, as indicated by IC analysis.

APPLICATIONS

- Gateway chemical to organophosphates and phosphazenes [1].
- Transformation of alcohols and acids into alkyl chlorides and acyl chlorides [2].
- Specific reagent for chlorinating aliphatic aldehydes and ketones to gem-dichlorides [3].
- Electrophilic substitution of active hydrogen compounds, for example, styrene [4].

18.1 INTRODUCTION

Phosphorus pentachloride (PCl_5) is a gateway chemical to a range of phosphorus-containing compounds [1] and a general chlorinating agent in organic synthesis. It is a stronger oxidizing agent than PCl_3, with chlorination often accompanied by oxidation. It reacts extremely violently with water, generating hydrochloric and phosphoric acids, and, hence, it is a sea-freight item.

Conventional laboratory preparations of phosphorus pentachloride generally involve the chlorination of white phosphorus in a suitable solvent (red phosphorus is largely insoluble), and are encumbered by the inconvenience of handling phosphorus. The laboratory procedure presented here is based on the thermal chlorination at 240°C of calcium phosphide, which is cheaply and commonly available due to its use as a fumigant and phosphine source. This reaction has been briefly mentioned by Moissan [5]. The information he provides is confined to the observation that at 100°C a reaction between chlorine and calcium phosphide proceeds with the evolution of bright light. Here, we find an almost quantitative yield, restricted only by the composition of technical calcium phosphide, which is generally very impure.

18.2 DISCUSSION

Phosphorus pentachloride in the solid state exists in the form of an ionic lattice of PCl^{4+} and PCl^{6-} ions. In the liquid state under pressure (at atmospheric pressure PCl_5 sublimes at 160°C) or in solution in nonpolar solvents, such as CCl_4, phosphorus pentachloride takes the form of covalent PCl_5 molecules. In polar solvents such as CH_3CN, it exists as a predominantly ionic solution [1]. In the gas phase, PCl_5 is partially reduced to PCl_3, so that gaseous PCl_5 is a three-component equilibrium:

$$PCl_5 \leftrightarrow PCl_3 + Cl_2, \quad K \equiv \frac{[Cl_2][PCl_3]}{[PCl_5]} = 0.042 \text{ at } 190°C. \qquad (18.1)$$

The moderate value of the equilibrium constant provides a simple method for interconverting between the +3 and +5 oxidation states of phosphorus in PCl_3 and PCl_5. A consequence of this is that purification of PCl_5 by sublimation has to be conducted in a chlorine atmosphere to reduce product loss.

In common with many phosphides, calcium phosphide is not strictly a stoichiometric compound. Various stoichiometries can be obtained by direct combination of the elements in different proportions, ranging from Ca_3P_2, CaP, to Ca_5P_8 [1], with impure formulations containing a mixture of these. The chemical properties of these phosphides vary with stoichiometry; thus, hydrolysis of Ca_3P_2 yields phosphine almost quantitatively, while hydrolysis of CaP generates diphosphine [1], a very unstable liquid at room temperature that ignites spontaneously in air.

The calcium phosphide used in the present experiment was purchased from Sigma-Aldrich [6]. The declared stoichiometry is Ca_3P_2, but no assay is available. The amount of phosphine generated by hydrolyzing with 1 M hydrochloric acid corresponds to only 40% Ca_3P_2. A significant amount of diphosphine was generated, evidenced by the spontaneous flammability of the liquid product after hydrolysis. No attempt was made to measure the equivalent CaP content, since the oxidation of the latter by chlorine is in all respects analogous to Ca_3P_2

$$Ca_xP_{2y} + (x+5y)Cl_2 \rightarrow xCaCl_2 + 2yPCl_5. \qquad (18.2)$$

Hydrolysis also revealed that the calcium phosphide contains a significant amount of inert matter, consisting predominantly of slag, carbon, transition metals, salts, etc. Despite the obviously low purity of this reagent, it was used in the present experiment because it represents a typical sample of the commercially available product. Purer phosphides are commercially available in small quantities; however, their extra expense is not justified in view of the high-purity PCl_5 obtained here, despite the impure nature of the reagent.

Although phosphorus is oxidized to PCl_5 in Reaction 18.2, the trivalent PCl_3 is mainly formed in the first stage of the present experiment. With a well-packed reactor tube, however, no observable PCl_3 exits the reaction zone (if the tube contains free space, some PCl_3 appears) despite the reactor temperature, 240°C, being well above the 76°C bp of PCl$_3$ (and 160°C sublimation temperature of PCl$_5$), and despite the complete absorption of chlorine within the reactor. It is thought that the reason for this is the basic character of calcium phosphide, so that as the PCl$_3$ is formed and vaporized from the reaction zone, being electrophilic, it forms an adduct of type Ca$_3$P$_2$.xPCl$_3$ as it comes into contact with unreacted Ca$_3$P$_2$ (an example of such a compound is mentioned by Urbain [7]). When the reaction zone reaches the end of the tube and the amount of Ca$_3$P$_2$ remaining becomes insufficient to bind all PCl$_3$ formed, the PCl$_3$ is liberated. Because now a substantial amount of chlorine is passing the reactor unreduced, the PCl$_3$ is oxidized to PCl$_5$, so that when all available phosphorus has been volatilized from the reactor zone, the product consists mostly of PCl$_5$.

18.3   EXPERIMENTAL

A borosilicate glass tube ~22 mm in diameter and ~20 cm long, with quickfit joints at both ends (19/26 minimum at the outlet end) is filled with 12.5 g (0.028 mol on Ca$_3$P$_2$ basis) of finely ground technical calcium phosphide on dry glass wool. Powdered Ca$_3$P$_2$ produced by an electric grinder can be rolled in glass wool and packed into the reactor tube through a powder funnel. Alternatively, coarse powder, which will not stick to glass wool, can be packed in layers 3–4 mm thick and separated by layers of glass wool. The latter arrangement results in a 5–10% loss of yield. Thicker layers lead to significant losses due to nonuniform propagation of the reaction along the length of the tube, forming pockets of solidified CaCl$_2$ that shield unreacted Ca$_3$P$_2$ from chlorine gas.

The outlet of the reaction tube is connected through a 90° bend to a two-neck, 100-mL, receiver flask cooled in ice water. The second neck of the receiver flask is fitted with a Liebig condenser connected in reflux mode, whose outlet is protected by a P$_2$O$_5$ tube and then vented through a bubbler. This arrangement is designed to condense any liquid PCl$_3$ formed, as well as the solid PCl$_5$, while avoiding blockage that could result if the condenser was connected directly to the reactor outlet. This part of the setup is shown in Figure 18.1.

The tube is placed inside a tube oven, and connected to a chlorine generator through a concentrated H$_2$SO$_4$ scrubber with a glass wool spray trap. The generator is charged with 60 g TCCA and 100 mL 15% HCl, sufficient to generate 0.5 mol chlorine. The reactor is purged with dry nitrogen through the chlorine generator to prevent any moisture present producing phosphine/diphosphine, which will react explosively with oxygen.

The tube oven temperature is raised to 240°C, and gas flow switched from nitrogen to chlorine, with the generator running at about 1 drop/sec. As the chlorine concentration builds up in the reactor tube, a dull red incandescence will be seen near the inlet, and start propagating slowly toward the outlet as the reagent is consumed. If the tube is well packed, there will be no output from the reactor at this stage, save for a small amount of gas passing the bubbler; in the alternative case, after a while, some liquid PCl$_3$ will accumulate in the receiver. After the reaction zone has traversed about ¾ of the tube, thick, white PCl$_5$ fumes will start effusing

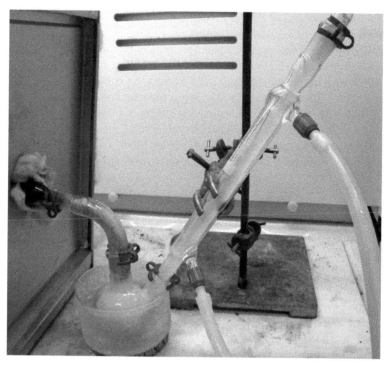

FIGURE 18.1 *A color version of this figure follows page 112.* Calcium phosphide heated to 240°C reacts with a dull red glow with chlorine gas, yielding PCl_3, which is further oxidized to PCl_5. After about three quarters of the stoichiometric amount of chlorine has been introduced, PCl_5 sublimes out of the reaction zone and is deposited as light yellow flakes on the walls of the receiver flask.

from the reactor and condensing in the flask, which will now be passing part of the chlorine as well.

After the reaction zone has traversed the entire tube, passage of chlorine is continued for about 10 min to ensure complete conversion of any PCl_3 formed to PCl_5, at which point the contents of the receiver should be completely solid. One obtains 12.3 g of light yellow PCl_5 flakes, which, if the reagent were pure Ca_3P_2, would represent a yield of 45%. However, as the reagent was found to contain only 40% Ca_3P_2 (see previous section), the remainder being inactive impurities and a small amount of CaP, we can conclude that the actual yield is much higher. Provided the glass wool contains no organic impurities, the product is already quite pure, with 100% subliming below 230°C. The chief impurity is PCl_3, whose presence cannot be altogether avoided due to the Equilibrium 18.1. Its content can be decreased by refluxing the product at 80°C in a chlorine flux; this is, however, accompanied by yield loss, due to some product escaping with the chlorine.

If desired, additional purification can be carried out by sublimation onto a cold finger at 230°C. In the present experiment, this sublimation was accompanied by an equilibrium redistribution between PCl_5 and PCl_3 corresponding to ~10 wt% PCl_3

in the final product, in accordance with the equilibrium constant in Equation 18.1. Phosphorus pentachloride must be stored in hermetically sealed containers. Schott screw-cap bottles do not provide an adequate seal, with ~10% product being converted to $POCl_3$ by reaction with atmospheric moisture in the course of several weeks.

At the completion of the experiment, the reagent tube should be emptied carefully due to the possibility of unreacted calcium phosphide powder remaining inside. This is best done by pouring some dilute HCl into the vertical tube and letting it percolate through it. Because some PH_3/P_2H_4 may be generated, this should be done inside a fume hood with no combustible material nearby. Pure PH_3 is nearly odorless; however, P_2H_4 has a strong odor reminiscent of arc-welding fumes (more so than garlic, to which it is often compared). The odor is not sufficiently irritating to reflect the extremely poisonous nature of these gases, which is comparable to hydrogen cyanide.

REFERENCES

1. Greenwood, N. N. and Earnshaw, A., *Chemistry of the Elements, 2nd ed.*, pp. 490–501. Oxford, U.K. : Butterworth-Heinemann, 1997.
2. Burks, J. E., Phosphorus(V) chloride. In *Encyclopedia of Reagents for Organic Synthesis*, edited by Paquette, L. New York: John Wiley & Sons, 2004.
3. Sandler, S. R. and Karo, W., *Organic Functional Group Preparations*, p. 134. New York–London: Academic Press, 1968.
4. Schmutzler, R., Styrilphosphonic dichloride. *Org. Synth. Coll.* 5: 1005–6, 1973.
5. Moissan, H., *The Electric Furnace.* (Trans. Lenher, V.) Easton, PA: The Chemical Publishing Co., 1904.
6. http://pubchem.ncbi.nlm.nih.gov/summary/summary.cgi?sid=24864973.
7. Urbain, E., Preparation of Chlorides of Phosphorus. U.S. Patent No. 1859543, May 24, 1932.

19 Phosphorus Oxychloride

SUMMARY

$POCl_3$ from P_2O_5 and NaCl

- Reaction commences at 270°C–300°C and is essentially complete by 420°C, producing phosphorus oxychloride in 85% yield based on reaction stoichiometry.
- Solid residue corresponds to a 1:2 Na_2O/P_2O_5 ratio rather than the 1:1 ratio of $NaPO_3$.
- Borosilicate glass is not attacked at the reaction temperature.
- $POCl_3$ is 99% pure by IC analysis.

$POCl_3$ from $Ca_3(PO_4)_2$, Cl_2, and C

- $Ca_3(PO_4)_2$–C mixture in a quartz reactor absorbs chlorine completely at 760°C with almost quantitative conversion of phosphorus to phosphorus chlorides.
- 80% phosphorus volatilization yields 87% $POCl_3$, and 13% PCl_3 (<0.5% PCl_5).
- Initial stages of reaction are accompanied by the complete conversion of $Ca_3(PO_4)_2$ to $Ca(PO_3)_2$, which can be used as an alternative starting material.
- A method is presented for producing active carbon suitable for the reaction.

APPLICATIONS [1]

- Essential reagent in the synthesis of phosphate esters $(RO)_3PO$ by reaction with alcohols in basic media. These find wide applicability as plasticizers, hydraulic fluids, flame retardants, and pesticides [2].
- Reaction with Grignard reagents is a convenient synthesis of phosphine oxides R_3PO.
- Forms an electrophilic intermediate with dimethylformamide in the Vilsmeier reaction for formylating aromatic rings and olefins with activating substituents.
- Complexes $AlCl_3$ in $PClO_3.AlCl_3$ (acts as a Lewis base), which is used to extract $AlCl_3$ in the workup of Friedel–Crafts reactions.
- Dehydrating agent, in the transformation of amides to nitriles, and specifically in Bischler–Napieralski cyclization.

19.1   INTRODUCTION

Phosphorus oxychloride is the mixed anhydride of phosphoric and hydrochloric acids, a corrosive liquid (bp 107°C) that fumes in moist air. In contact with cold water, it initially collects in a dense immiscible layer. However, after a short delay, a very vigorous reaction commences, accelerated by its exothermicity, producing phosphoric and hydrochloric acids. Its vapors are lachrymatory and poisonous, with a possible latent physiological action [2] similar to that of phosgene, with which it shares a structural and functional similarity. Hence, phosphorus oxychloride is a sea-freight item. Industrially, phosphorus oxychloride is manufactured by either the oxidation of PCl_3 with oxygen, or with chlorine in the presence of P_2O_5. Laboratory methods rely on either the partial hydrolysis of PCl_5, or the reaction between P_2O_5 and PCl_5 [1,2]. Because phosphorus chlorides are themselves sea-freight items, two preparations from easily available chemicals are presented here.

The first preparation is based on the reaction between P_2O_5 and NaCl, which was first described by Kolbe [3] without a detailed description of reaction conditions. It was also reported by Tarbutton et al. [4], where a steel autoclave was used at 250°C–400°C. This corresponds to pressures of several MPa, and the steel reactor reduces some of the $POCl_3$ formed to PCl_3. Neither article provides details of yield or stoichiometry. In the experiment presented here, the reaction commences at 270°C–300°C and is essentially complete by 420°C. The stoichiometry differs from what might have been naively expected. While it is clear a priori that the triphosphate Na_3PO_4 is too basic to be the final product, one might have expected a residue with a 1:1 Na_2O/P_2O_5 ratio, namely, the reaction

$$2P_2O_5 + 3NaCl \rightarrow 3NaPO_3 + POCl_3. \tag{19.1}$$

Instead, a ratio approaching 1:2 Na_2O/P_2O_5 has been found in all cases:

$$7P_2O_5 + 6NaCl \rightarrow 3Na_2P_4O_{11} + 2POCl_3. \tag{19.2}$$

Based on this stoichiometry, a yield of 76–86% is obtained. The reaction can be conducted in ordinary borosilicate glass, which is not attacked by dry P_2O_5.

The second preparation is based on the reduction of the cheaper and more readily available tricalcium phosphate $Ca_3(PO_4)_2$ with carbon in the presence of chlorine at 760°C in a quartz reactor. This process is described on an industrial scale with the following stoichiometry [5]:

$$Ca_3(PO_4)_2 + 6C + 6Cl_2 \rightarrow 3CaCl_2 + 6CO + 2POCl_3, \tag{19.3}$$

$$Ca_3(PO_4)_2 + 8C + (6Cl_2, 8Cl_2) \rightarrow 3CaCl_2 + 6CO + (2PCl_3, 2PCl_5). \tag{19.4}$$

Phosphorous volatilization of 97% is reported in excess chlorine, but no information is presented on the amount of chlorine required or the ratio of PCl_3, PCl_5, and $POCl_3$ in the final product.

The present experiment demonstrates that chlorine generated at 0.35 mol/h is absorbed completely even by thin layers of tricalcium phosphate–carbon mixtures at 760°C, enabling almost complete conversion to the phosphorus chlorides. The product consists predominantly of $POCl_3$ (87%), with the remaining 13% consisting almost entirely of PCl_3 (<0.5% PCl_5) if the reaction is halted at about 80% phosphorus volatilization, which corresponds to the first appearance of PCl_5 (white sublimate) in the receiver.

Two further findings are worthy of note. First, the first 2 mol equivalents of chlorine are consumed without any liquid product appearing in the receiver. This is due to the basicity of tricalcium phosphate, so that the $POCl_3$–PCl_3 reacts with excess $Ca_3(PO_4)_2$ to yield a more acidic calcium phosphate with a smaller CaO-to-P_2O_5 ratio. Second, the amount of chlorine required is about a factor of two in excess of Reaction 19.3, due to phosgene rather than carbon monoxide being the major gaseous by-product of the reaction.

19.2 DISCUSSION

19.2.1 REACTION BETWEEN P_2O_5 AND NACL

The present preparation leaves a residue with the nominal composition $Na_2P_4O_{11}$ in the form of a hard solid mass, which dissolves very slowly in water even under basic conditions. The dissolution process is uniform, and the heat released is much less than would be expected from an equivalent mixture of P_2O_5 and $NaPO_3$, showing that the 2:1 P_2O_5/Na_2O stoichiometry does not contain an unreacted P_2O_5 component, but rather represents an extended polymeric structure. The possibility of a significant P_2O_5 component dissolving in $POCl_3$ with the formation of a polymeric compound of type $(POCl_2)_n$ is discounted in the present experiment, since this substance decomposes above 150°C [6].

Phosphorus oxychloride is also formed by the dissolution of P_2O_5 in HCl, which is an analog of Reaction 19.1 in acidic media [7]:

$$2P_2O_5 + 3HCl \rightarrow 3HPO_3 + POCl_3. \tag{19.5}$$

This interaction, however, is exceedingly slow under standard conditions in typical laboratory apparatus, with HCl absorbed at a rate of about 1 mL/h [7], while operation at higher temperatures is complicated by the action of HPO_3 on glass.

19.2.2 REACTION BETWEEN $CA_3(PO_4)_2$, CARBON, AND CHLORINE

This reaction is analogous to the well-known two-stage industrial preparation of PCl_3 and $POCl_3$ by reduction of tricalcium phosphate with carbon at 1350°C, followed by chlorination of the phosphorus at 100°C–200°C [8]

$$Ca_3(PO_4)_2 + 5C \rightarrow 3CaO + 5CO + P_2 \qquad \Delta G_{730°C} = +517 \text{ kJ/mol}, \tag{19.6}$$
$$\Delta G_{1330°C} = -370 \text{ kJ/mol},$$

$$P_2 + 3Cl_2 \rightarrow 2PCl_3, \qquad \Delta G_{730°C} = -460 \text{ kJ/mol}, \tag{19.7}$$

$$3PCl_3 + 3Cl_2 + P_2O_5 \rightarrow 5POCl_3. \tag{19.8}$$

While phosphorus formation is highly endothermic, Reaction 19.6 proceeds at high temperatures due to the positive entropy change associated with the liberation of 6 mol of gaseous products. While this shows that at lower temperatures substantial reaction is precluded by thermodynamic rather than just kinetic considerations, it opens the possibility of coupling the negative ΔG of chlorination with the positive ΔG of phosphorus formation, enabling the reaction to proceed at much lower temperatures. Confirmation of this has been reported by Vivian [5], with the overall process represented by Reaction 19.3. The reaction may additionally proceed by the formation of chlorinated carbon compounds, which subsequently react with tricalcium phosphate, by analogy with the carbochlorination of transition metal oxides reported at similar temperatures in Reference 9.

In the present experiment, the $Ca_3(PO_4)_2$–C fully absorbs chlorine introduced at 0.35 mol/h with a reagent cross section of 3.8 cm², and thickness changing in the course of the reaction from 20 to 5 cm, corresponding to space velocities from 0.11 to 0.43 s⁻¹. When chlorine is merely passed over the solid reagents at 760°C, about 80% of it reacts based on gas composition analysis. This reactivity, combined with the almost complete conversion of the reagents, strongly suggests that the individual reaction steps do not involve the interaction of two solid phases, since this normally entails only partial conversion (see Chapter 5). This is in agreement with Jacob and Reynolds [8], where it was suggested that the initial step in Reaction 19.6 is the unimolecular decomposition of $Ca_3(PO_4)_2$ to the volatile P_2O_5 in the rate-determining step, followed by a rapid reduction of P_2O_5 by carbon. In the present experiment, the presence of PCl_3, where phosphorus is in a lower oxidation state than in $Ca_3(PO_4)_2$, suggests reduction of P_2O_5 by carbon followed by oxidation of the resulting phosphorus by chlorine, Reaction 19.7, as the method of PCl_3 formation. The alternative mechanism involving the action of volatile chlorocarbons on $Ca_3(PO_4)_2$ does not appear to have sufficient reducing power to produce PCl_3 directly. Most of the PCl_3 formed then reacts with the P_2O_5 component of $Ca_3(PO_4)_2$ according to Reaction 19.8, resulting in 87% $POCl_3$ in the final product.

As mentioned in the introduction, the reaction initially proceeds with complete absorption of chlorine, but no evolution of liquid products from the reactor. Thus, with 73.6 g (0.23 mol) of 96% $Ca_3(PO_4)_2$ and excess carbon, 0.45 mol of Cl_2 are completely absorbed before the first drops of $POCl_3$–PCl_3 exit the reactor. This corresponds to a phosphate-to-chlorine mole ratio of 1:2, suggesting the following reaction:

$$Ca_3(PO_4)_2 + 2C + 2Cl_2 \rightarrow 2CaCl_2 + Ca(PO_3)_2 + 2CO, \tag{19.9}$$

which alters the CaO-to-P_2O_5 mole ratio from 3:1 in tricalcium phosphate to 1:1 in the more acidic calcium metaphosphate. While both reactions, Reaction 19.9 and Reaction 19.3, can occur simultaneously, the $POCl_3$ formed in Reaction 19.3 does not exit the reactor tube, as it is completely absorbed by the basic $Ca_3(PO_4)_2$:

$$2Ca_3(PO_4)_2 + 2POCl_3 \rightarrow 3CaCl_2 + 3Ca(PO_3)_2, \tag{19.10}$$

leading to Reaction 19.9 as the overall result, until all tricalcium phosphate has been converted to the metaphosphate. Reaction 19.9 can be avoided by starting directly from the calcium metaphosphate. This can be prepared by neutralizing tricalcium phosphate with a stoichiometric amount of sulfuric acid to form the soluble acid phosphate $Ca(H_2PO_4)_2$, which is then separated from $CaSO_4$ by filtration, and dehydrated to the metaphosphate at red heat. This process was used in the earliest methods of phosphorus production, where only the excess P_2O_5 in calcium metaphosphate is reduced to phosphorus at red heat with the metaphosphate being converted to tricalcium phosphate in the process.

Calcium metaphosphate reacts with carbon and chlorine to form $POCl_3$ in the following reactions:

$$Ca(PO_3)_2 + 4Cl_2 + (4C, 2C) \rightarrow CaCl_2 + 2POCl_3 + (4CO, 2CO_2), \quad (19.11)$$

$$Ca(PO_3)_2 + 8Cl_2 + 4C \rightarrow CaCl_2 + 2POCl_3 + 4COCl_2. \quad (19.12)$$

Oxidation of carbon to CO_2 is not observed in the reduction of phosphorus [8], but occurs in the chlorination Reaction 19.3 [2], while the formation of phosgene is thermodynamically unfavorable above about 530°C. Nevertheless, in the present experiment, the ratio of chlorine consumed to $POCl_3$ formed is much closer to that of Reaction 19.12 than to the other possibilities. Thus, formation of 51.2 g of product consumed 2.1 mol of Cl_2, compared to 1.8 mol predicted by Reaction 19.10 and Reaction 19.12, with the difference likely attributable to the formation of inert chlorocarbons, as no free chlorine passed the reactor. Hydrolysis of product gases in basic solution gave a Cl_2-to-CO_2 ratio close to 1, also in accordance with phosgene being a major reaction product. The mole ratio of bound Cl_2 to CO in the gaseous reaction products was measured at about 2:1. The weight change of the solid residue for 51.2 and 25 g of $POCl_3/PCl_3$ produced was measured at −15 g and 0 g, respectively, while Reaction 19.10, the first step in Reaction 19.11, and Reaction 19.12 predict −8.3 g and +4.7 g for the two cases. It is clear then that in addition to processes 19.10–19.12, a substantial amount of carbon is lost in the formation of chlorocarbon by-products, justifying the excess suggested by Vivian [5] and used in the present preparation.

19.3 EXPERIMENTAL

19.3.1 PREPARATION FROM P_2O_5

The best reaction vessel for this preparation is a test tube with a quickfit joint, a capacity of at least 100 mL, and at least 22 mm in diameter. If this is not available, a 100-mL, single-neck flask can be used instead with a slight loss of yield. Scaling the reaction requires proportionally larger vessels.

First, 35.1 g (0.25 mol) of phosphorus pentoxide is rapidly introduced into the reaction vessel, ensuring that as little atmospheric moisture as possible reacts with the P_2O_5, as it subsequently generates HCl and HPO_3, which attacks the glass. The transfer is best done by inverting a powder funnel inserted into the reaction vessel, over the mouth of the reagent bottle, and tilting until approximately the right amount of reagent is transferred into the flask. Next, 34 g (0.58 mol) of finely ground and thoroughly desiccated (1 h at 250°C) NaCl is added, and the reagents thoroughly mixed. The approximately twofold excess of NaCl improves the yield by about 10%. Finally, about 10 g NaCl is poured in a layer on top of the mixture, which serves to convert unreacted P_2O_5 subliming from the reaction zone (bp 360°C). The reaction vessel is placed in an air oven and connected to a bend leading through a Liebig condenser to a receiver flask immersed in cold water. The outlet from the flask is vented through a $CaCl_2$ protection tube with a bubbler optionally attached for observation of reaction progress.

The oven is rapidly heated to 270°C, from where the temperature is raised more slowly to 450°C at about 65°C/h. A distillation conducted more rapidly than this will serve both to reduce yield by excessive sublimation of P_2O_5, and increase the possibility of the reagent vessel bursting due to excessive buildup of HCl and $POCl_3$ pressure inside the reactor, as these do not have sufficient time to percolate through the viscous sodium metaphosphate medium (the reagents liquefy during the reaction). About 8.2 g (0.053 mol) $POCl_3$ collects in the receiver at the end of the reaction while the reactor tube loses 9.3 g (the difference is due to losses in the distillation setup), corresponding to a yield of 76% (86% with respect to reactor weight loss) based on P_2O_5.

19.3.2   PREPARATION OF SUGAR CHARCOAL

This form of carbon is particularly suitable for gaseous thermal reduction due to its porosity, purity, and low ash content. Marvin et al. [10] describes a small-scale synthesis (12 g) in a 1-L container, while here we describe a preparation of 300 g in a slightly larger reactor volume.

Begin by placing 1500 g (4.4 mol) of food-grade sugar, $C_{12}H_{22}O_{11}$, in a high-form, 3-L beaker and heated inside an air oven at 200°C for several hours. When the sucrose has melted, substantially dehydrated, and charred, the temperature is raised to 350°C in the course of several hours, during which the volume of the contents increases severalfold and a matrix of black, low-density carbon is formed. If the mixture ignites during this period, a lid is placed on top of the beaker until it is extinguished. The carbon is finally allowed to cool, ground in a mortar, and calcined at 600°C for 1 h. The yield of black porous charcoal is 300 g, or 47% of the theoretical yield.

19.3.3   PREPARATION FROM $CA_3(PO_4)_2$

This reaction must be carried out in a quartz tube or U-tube heated in either a tube oven or a box oven, respectively. Because the calcium chloride product attacks quartz in depth and also expands substantially on cooling, the reagents are best contained in a sacrificial inner glass tube, which can be either quartz or ceramic. Borosilicate glass also can be used provided the maximum temperature is not raised above 720°C.

First, 65.5 g (0.2 mol) of 96% $Ca_3(PO_4)_2$ is ground to a fine powder mixed with 22 g of sugar charcoal and placed inside a glass tube about 250 mm long and 22 mm in diameter. The ends of the tube are lightly blocked with glass wool, and the tube is centered inside a quartz reactor tube using tightly wedged glass wool padding at both ends to minimize gas leakage outside the inner tube. If a borosilicate glass tube is used, the inner tube also must be supported by ceramic wool padding underneath.

The preparation can be carried out as well in a standard box oven using a quartz U-tube with quickfit joints (Figure 19.1). In this case, two inner tubes packed with reagent and lightly blocked at the lower end with ceramic wool are inserted into both straight legs of the U-tube and supported by ceramic wool. A gas inlet tube is attached to one end of the quartz reactor, while the other end is connected to a receiver flask as in the previous preparation. Chlorine gas is generated as described in the PCl_5 preparation procedure.

Initially, the reactor is disconnected from the receiver, while a slow current of dry nitrogen is introduced through the chlorine generator head, and the oven temperature is raised to 760°C. This arrangement dehydrates tricalcium phosphate over several hours. When water vapor is no longer evolved, which can be checked by attaching a

FIGURE 19.1 *A color version of this figure follows page 112.* A mixture of powdered carbon and $Ca_3(PO_4)_2$ is chlorinated at 760°C. Initially, $Ca_3(PO_4)_2$ is converted to $Ca(PO_3)_2$. Subsequently, $POCl_3$ forms and collects in the receiver. The gaseous products include CO and $COCl_2$; hence, the preparation is conducted inside a fume hood.

short length of cold glass tube to the reactor outlet and observing that no condensation forms over a period of several minutes, the receiver is reattached to the reactor
and a flow of chlorine commenced at a rate corresponding to about 1 drop HCl per
5 sec. The flow of chlorine is accompanied by vigorous bubbling of outlet gases containing no chlorine (color), while the last remnants of water (<0.1 g) exit the tube as
a fine mist depositing in the distillation bend.

After the chlorine generator has consumed about 80 mL of HCl, a transparent
liquid begins collecting in the receiver at a rate of about 1 drop/sec. The experiment is continued until a total of 420 mL HCl has been consumed in the chlorine
generator (5–6 h), at which point there should appear the first signs of chlorine gas
exiting the reactor in the form of a fine PCl_5 mist on the receiver flask walls. Gas
generation is then stopped, the reactor cooled, and flushed with nitrogen. The 51.2 g
of liquid product was fractionated, yielding two fractions, one boiling at 76°C (PCl_3),
the other at 107°C ($POCl_3$). Higher boiling products (PCl_5) amounted to less than
1%. The yield of phosphorus chlorides to the point where Cl_2 appears in the reaction
products is thus 83%, based on tricalcium phosphate in Reaction 19.3 (100% yield
from 63 g $Ca_3(PO_4)_2$ is 62 g $POCl_3$, or 56 g PCl_3).

19.3.4 Determination of PCl_3

This was established by hydrolysis of the products followed by iodometric titration.
The former generates phosphorous acid, which is oxidized to phosphoric acid with
excess iodine and back-titrated with thiosulfate.

$$PCl_3 + 3H_2O \rightarrow H_3PO_3 + 3HCl, \tag{19.13}$$

$$H_3PO_3 + I_2 + H_2O \rightarrow H_3PO_4 + 2HI. \tag{19.14}$$

The hydrolysis is carried out by quickly adding a large excess of chilled water to the
test sample (if only a small amount of water is added, localized heating can lead to
some thermal decomposition of H_3PO_3 to PH_3 while a substantial amount of $POCl_3$
hydrolysis product is deposited as the little-soluble $(HPO_3)_n$ gel). Excess iodine is
added in the form of an I_2 solution in KI (2 g KI per 1 g I_2 dissolved initially in a
minimum amount of water) with the redox Reaction 19.14 carried out in excess phosphate buffer formed by adjusting a solution of $NaOH/H_3PO_4$ to 7.3 pH [11,12].

Phosphorus oxychloride is soluble in nonpolar compounds and hydrolyzes in
water after a small delay, generating HCl and HPO_3. Because it is structurally
and functionally similar to phosgene, it is extremely toxic by inhalation (TLA of
0.1 ppm), with similar symptoms of delayed pulmonary edema. Being a liquid, it
normally presents much lower vapor pressures than phosgene; however, this pro-
tective feature is diminished in the present experiment where its partial pressure
is above atmospheric in the reactor. It reacts far more vigorously with water than
phosgene, thus providing more effective warning by irritation of the upper respira-
tory tract. However, its odor threshold is similar to that of phosgene (0.5 ppm). At
this concentration, it has a sharp odor resembling HCl, but takes a few seconds

to register. Unlike HCl, after the odor is detected, the sensation can persist for hours, being accompanied by a more general irritation than an equivalent amount of HCl produces. Thus, the present experiment must be conducted in an effective fume hood.

REFERENCES

1. Meier, M. S., and Ruder, S. M., Malona, J. A., and Frontier, A. J., Phosphorus oxychloride. In *Encyclopedia of Reagents for Organic Synthesis*, edited by Paquette, L. New York: John Wiley & Sons, 2004.
2. Fee, D. C., Gard, D. R., and Yang, C.-H., Phosphorus compounds. In *Kirk-Othmer Encyclopedia of Chemical Technology, 5th ed.*, Vol. 19. New York: John Wiley & Sons, 2006.
3. Kolbe, H., Verhalten der Wasserfreien Phosphorsaure. *Eur. J. Org. Chem. (Ann. Chem.)* 113(2): 240, 1860.
4. Tarbutton, G, Egan, E. P., and Frary, S. G., Phosphorus-halogen compounds from phosphorus pentoxide and halides. Properties of phosphorus trifluoride and phosphorus oxyfluoride. *J. Am. Chem. Soc.* 63 (7), 1782–1789, 1941.
5. Vivian, R. E., Process of Extracting Phosphorus Content from Phosphorus Containing Materials. U.S. Patent No. 1926072, Sep. 12, 1933.
6. Huntly, G. N., Action of phosphoryl chloride on phosphorus pentoxide. *J. Chem. Soc., Trans.* 59: 202–8, 1891.
7. Bailey, G. H. and Fowler, G. J., Some reactions of the halogen acids. *J. Chem. Soc., Trans.* 53: 755–61, 1888.
8. Jacob, K. D. and Reynolds, D. S., Reduction of tricalcium phosphate by carbon. *Ind. Eng. Chem.* 20(11): 1204–10, 1928.
9. Landsberg, A., Wilson, R. D., and Burns, W., Conditions affecting the formation of chlorinated carbon compounds during carbochlorination. *Met. Trans. B*, 19B(3): 477–82, 1988.
10. Marvin, G. G., Booth, H. S., and Dolance, A., Sugar charcoal. *Inorg. Synth.* 2: 74–75, 1946.
11. Van Name, R. G. and Huff, W. J., The estimation of phosphorous, hypophosphoric and phosphoric acids in mixture. *Am. J. Sci.* 45: 91–102, 1918.
12. Joint FAO/WHO Expert Committee on Food Additives. Compendium of Food Additive Specifications, Addendum 12, p. 22. Paper presented at the 63rd meeting of FAO/WHO, June 8–17, 2004, Geneva, Switzerland.

20 Sulfur Trioxide and Oleum

SUMMARY

- SO$_3$/oleum is prepared by the pyrolysis of NaHSO$_4$ to a maximum temperature of 820°C.
- A 500°C cut gives 86% oleum at 90% yield. A 580°C cut gives 94% oleum at 65% yield.
- 100% volatilization of H$_2$SO$_4$ from NaHSO$_4$ shows that the residue is entirely Na$_2$SO$_4$.
- The reaction can be conducted in an unprotected quartz tube. Na$_2$SO$_4$ does not attack quartz at the reaction temperature.

APPLICATIONS

- Sulfonation and nitration of unreactive or deactivated aromatics, for example, benzene, anthraquinone, or m-dinitrobenzene [1].
- Formation of stable adducts with Lewis bases, for example, pyridine and quinoline, used for separation from reaction mixtures (see Chapter 21).
- Preparation of other strong Lewis acids, for example, nitryl chloride (NO$_2$Cl), used to chloronitrate olefins [2], chlorosulfonic acid, thionyl chloride (see Chapter 21), and peroxysulfuric acid [3].

20.1 INTRODUCTION

Solid sulfur trioxide has a trimeric ice-like structure: γ-SO$_3$ [3]; however, a water content as low as 10^{-3} mol% alters this substantially, so that sulfur trioxide is most commonly encountered as the needle-like α-SO$_3$ or β-SO$_3$ phases, consisting of large polymeric chains HO(SO$_3$)$_n$H, which are also typical of high-concentration oleum. Oleum is most frequently produced as a SO$_3$/H$_2$O composition liquid at room temperature, corresponding to 0–30% or 65–75% free SO$_3$, with the latter containing only ~5 wt% water. The liquid state frequently confers better mixing and increased reactivity, as well as making the reagent easier to manipulate (see Chapter 21).

Sulfur trioxide is a sufficiently strong oxidizing agent to oxidize carbon at room temperature, and an aggressive dehydrating agent, reacting violently with water and alcohols. Even low-pressure vapor instantly carbonizes most organic materials, including polyethylene and silicone grease (see section 20.3). Thus, it is a sea-freight item and is generally quite expensive.

The large quantities of sulfur trioxide and sulfuric acid used in industry are produced by oxidizing SO_2 in air on a vanadium pentoxide or platinum catalyst, but are too involved to be useful in the laboratory. A well-known laboratory method for preparing sulfur trioxide is based on the dehydration of sulfuric acid by P_2O_5 at 200°C–300°C [4]. This method is quite inefficient because the reaction is heterogeneous and requires a large excess of P_2O_5, and metaphosphoric acid rather than orthophosphoric acid is formed as the end product

$$H_2SO_4 + P_2O_5 \rightarrow 2HPO_3 + SO_3. \tag{20.1}$$

Moreover, it is difficult to find a suitable laboratory reaction vessel because the metaphosphoric acid attacks glass at the temperature of the reaction.

Another method cited in the literature is the thermal decomposition of sulfates (with the exception of alkali metal sulfates) and pyrosulfates, with sodium pyrosulfate appearing particularly suitable due to one of the lowest quoted decomposition temperatures [4]:

$$2NaHSO_4 \rightarrow Na_2S_2O_7 + H_2O, \qquad T_{dec} = 310°C \tag{20.2}$$

$$Na_2S_2O_7 \rightarrow Na_2SO_4 + SO_3. \qquad T_{dec} = 450°C \tag{20.3}$$

However, apart from the decomposition temperatures, no details are given of the experimental setup, dependence of H_2O and SO_3 volatilization on temperature or purity of the product.

The present experiment shows that the usual description of the decomposition as the two-stage Reaction 20.2–Reaction 20.3 is an idealization with the pyrosulfate stoichiometry not being achieved at any given temperature. Rather, the processes are intermixed, so that the reagent is better regarded as a particular composition of $Na_2S_2O_7/Na_2SO_4/H_2O$ or, equivalently, $Na_2SO_4/SO_3/H_2O$, with the ratios varying with temperature. The temperature region over which sodium pyrosulfate decomposes significantly is here found to be 260°C–820°C, much wider than quoted above, with more than 60% of SO_3 volatilized above 580°C. However, while both H_2O and SO_3 volatilize in the low temperature range, nearly pure SO_3 is evolved at the other temperature extreme, making this a useful method for preparing sulfur trioxide.

In the present experiment, a 580°C cut for the SO_3 fraction gives a 65% yield of SO_3/oleum with strength >90% determined by distillation. One hundred percent volatilization of H_2SO_4 from $NaHSO_4$ is obtained at a maximum temperature of 820°C, so that the residue consists entirely of sodium sulfate. Even though SO_3 is thermodynamically unstable with respect to SO_2 at that temperature [5], its decomposition is kinetically hindered in the absence of metal catalysts, so that although SO_2 is evident in the product collected at all temperatures, this is well below its solubility limit in H_2SO_4 (about 5% H_2SO_4 [5]) and no gas evolution was observed at any stage.

The reaction generally proceeds as follows: $NaHSO_4$ melts at 185°C–195°C, with H_2O evolution commencing at about 220°C, while white fumes start to evolve at 260°C, indicating the presence of SO_3. From 260°C to 420°C, the vapor composition is a fairly constant 16% H_2SO_4, with 75% of the water content evaporating in this range. The mp of pure $Na_2S_2O_7$ is quoted at 420°C; however, the reagent does not solidify up to the nominal decomposition point of $Na_2S_2O_7$ at 450°C, since pure $Na_2S_2O_7$ stoichiometry is never present. A result of practical significance in the present experiment is that, unlike the case with $CaCl_2$ (see Chapter 19), Na_2SO_4, which is solid at the maximum reaction temperature, does not attack quartz or expand on cooling, making it possible to conduct the reaction inside an unprotected quartz tube with distillation side arm.

20.2 DISCUSSION

20.2.1 THE SO_3–H_2O SYSTEM

The chemical composition of this system is a complex mixture of interacting components, including the neutral species H_2SO_4, SO_3, H_2O, $H_2S_2O_7$ and ionic species H_3O^+, HSO_4^-, $H_3SO_4^+$, and $HS_2O_7^-$, SO_4^{2-} [3]. While the contribution of the various species changes with stoichiometry, the latter is simply described by

$$a\ SO_3 + b\ H_2O \equiv a\ H_2SO_4 + (b - a)\ H_2O = b\ H_2SO_4 + (a - b)\ SO_3. \quad (20.4)$$

Mixtures with $a \ll b$ are dilute sulfuric acid, $a = b$ is 100% sulfuric acid, while $a > b$ corresponds to oleum, with the amount of free SO_3 given by $(a - b)/(a + 0.22\ b)$. Oleum has a peculiar nonmonotonic solid–liquid phase transition with two minima (Figure 20.1) [5]. Thus, it is a solid at room temperature in the interval 30–60% SO_3, and again at >75% SO_3, so that one cannot determine its strength on the grounds of a solid phase being present at room temperature. Oleum melting point determination

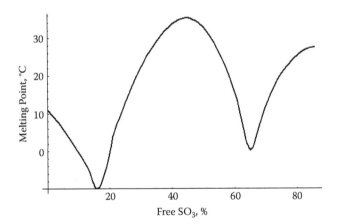

FIGURE 20.1 The melting point of oleum as a function of free SO_3 content. The extremes correspond to 100% H_2SO_4 and pure SO_3 (γ-phase), respectively. In between, oleum shows a nonmonotonic behavior with both weak 0–30% oleum and strong 65–75% oleum being liquid at room temperature.

is very inaccurate due to hysteresis; hence, the best method to determine strength is through a titration of the homogeneous product or gravimetric evaporation of SO_3.

Returning to the foregoing equation, $b = 0$ corresponds to pure sulfur trioxide, a white deliquescent fuming solid, which in the γ-phase has a melting point of 16°C and boiling point of 44°C. As mentioned in the introduction, sulfur trioxide easily polymerizes (enhanced by H_2O, inhibited by P_2O_5) with the three modifications α-, β-, and γ-SO_3, melting at 62°C, 32°C, and 16°C, respectively. This very narrow range of temperatures where SO_3 is a liquid is a source of problems in its preparation.

On evaporating sulfuric acid with $b > a$, the fraction of SO_3 in the distillate progressively increases until the maximum boiling point of 337°C of the 98% H_2SO_4 azeotrope is reached. After this, no further concentration in the SO_3 component can be obtained. Hence, distillation cannot be used to prepare oleum/SO_3.

20.2.2   THE SO_3–H_2O–NA_2SO_4 SYSTEM

Oleum can be obtained by the pyrolysis of $NaHSO_4$, which in the fluid state is equivalent to an equimolar mixture of H_2SO_4 and Na_2SO_4:

$$2NaHSO_4 \equiv H_2SO_4 + Na_2SO_4. \qquad (20.5)$$

Sodium sulfate is weakly basic, being the salt of the semiweak acid $NaHSO_4$, and we thus see from Equation 20.5 that Na_2SO_4 significantly increases the pH of H_2SO_4 solutions, with an equimolar mixture raising the pH of 1 mol/L H_2SO_4 from pH < 0 to pH ~ 1 (buffering). The weak basicity of Na_2SO_4 also qualitatively explains the low decomposition temperature of the pyrosulfate (compare to $CaSO_4 \rightarrow CaO + SO_3$ at T ~ 1050°C, CaO being a much stronger base).

Equation 20.5 turns out to be a realizable reaction in practice. The present experiment shows that H_2SO_4 can be distilled from $NaHSO_4$ with almost 100% efficiency. While collecting the distillate as a single fraction gives 100% H_2SO_4, unlike the SO_3/H_2O system, introducing a cut in the distillate fraction yields oleum, with the lower bp fraction being therefore a correspondingly weaker acid (because the combined fractions must give 100% H_2SO_4). *We can say that the addition of Na_2SO_4 to H_2SO_4 breaks the SO_3–H_2O azeotrope.*

Figure 20.2 shows data from the present experiment for the percentage of H_2O and SO_3 volatilized in a slow distillation as a function of temperature. From this, it is seen that the separation of the two volatile components is reasonably efficient, so that at 500°C 90% H_2O and only 10% SO_3 has volatilized, corresponding to an oleum strength of 86% free SO_3 at 90% yield in the upper cut. A cut at 580°C gives 94% oleum at 65% yield.

20.3   EXPERIMENTAL

The best reaction vessel for this experiment is a quartz test tube with integral distillation side arm. Figure 20.3 shows a such tube of 30 mm diameter, containing the

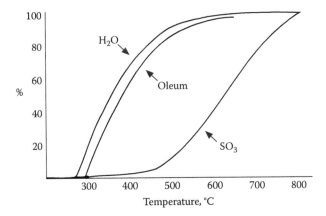

FIGURE 20.2   Fraction of H_2O and SO_3 volatilized in $NaHSO_4$ pyrolysis, as a function of temperature. The temperature overlap shows that an intermediate with exact $Na_2S_2O_7$ stoichiometry is not achieved. A third curve plots the oleum strength for a fraction collected between the plotted temperature and the 820°C maximum.

Na_2SO_4 residue at the end of the reaction. If a tube with integral side arm is not available, an ordinary quartz test tube can be substituted and the side arm attached to the ground glass joint using high-temperature grease. Because the maximum distillate temperature is about 340°C, there is the possibility of the joint seizing due to carbonization of the grease, and it should be separated while still hot.

The tube is filled with 170 g $NaHSO_4$ (about ¾ full), technical grade being acceptable for most purposes since the distillation purifies the product. An excessive content of organic matter or transition metal salts should be avoided, as the former results in a large amount of carboneous material distilling with the product while the latter catalyzes the decomposition of SO_3 to SO_2 and O_2, which is thermodynamically favored at the reaction temperatures. Technical-grade $NaHSO_4$ frequently contains a few percent water (the specimen in the present experiment contained 2%), while the stoichiometry also varies about Equation 20.5 by a few percent. Such variations are of no practical significance here.

The tube is placed inside a box oven inclined at about 20°, with the side arm connected to a water-cooled Liebig condenser and receiver, so that the product distills downward. The section of the quartz tube protruding from the oven is thermally insulated with glass wool to prevent condensation there. The oven temperature is rapidly raised to 180°C, and from there to the cut temperature (between about 450°C and 580°C) at no more than 150°C/h to avoid excessive evolution of vapor, which can splash solid product into the side arm. When the cut temperature is reached, the Liebig condenser and receiver are replaced by a 200 to 300-mm air condenser and a 1-L, two-neck flask, with about half its surface immersed in an ice-water bath. The large cooled surface area of the flask condenses the SO_3 vapor, as most product transfer occurs by sublimation. The other neck of the flask is connected to an ice water-cooled condenser that traps most of the remaining SO_3, and the outlet is vented through a $CaCl_2$ or P_2O_5 protection tube. As the oven temperature rises, the

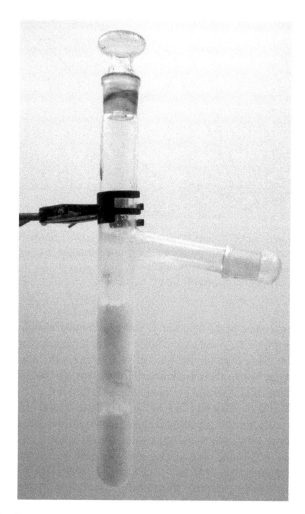

FIGURE 20.3   Quartz test tube with distillation side arm containing the Na$_2$SO$_4$ product at the end of the experiment.

flask walls become coated with a layer of SO$_3$ needles, while some liquid SO$_3$/oleum, which can assume a black coloration from interaction with traces of organic material in the reagent or the sealing grease, flows down the air condenser into the bottom of the flask (Figure 20.4).

The reactor is held at the ultimate temperature of 820°C until no further product sublimes into the receiver (about 30 min), after which the receiver is disconnected and stoppered, and the reactor cooled. The product can be freed from impurities by distilling the SO$_3$ into a fresh flask connected to the first by means of a wide 120°–180° bend. The top surface of the distillation flask is insulated and its bottom heated either on an oil bath or an air oven (or with air gun), with a large part of the receiver flask cooled in an ice-water bath. As before, a moisture protection tube must be used. The distillation should not be done too rapidly (about ½ h), and the SO$_3$ can

FIGURE 20.4 *A color version of this figure follows page 112.* The fraction above 500°C from NaHSO$_4$ pyrolysis. Liquid oleum, colored black from carbonization of silicon grease, collects in the receiver below 580°C; above this temperature, nearly pure SO$_3$ deposits on the chilled flask walls by sublimation from the reactor.

take a while to solidify due to hysteresis. Using a cut at 580°C, the yield is 39.8 g SO$_3$ or 68% based on the free SO$_3$ content of NaHSO$_4$.

Oleum of any required concentration can be formed from this product by slowly running in the calculated amount of 98% H$_2$SO$_4$ (the reaction with water or dilute H$_2$SO$_4$ is too energetic and produces copious fumes) into a flask containing stirred molten SO$_3$ on a fixed-temperature paraffin bath.

Unlike SO$_2$, SO$_3$ possesses no warning odor; however, its vapors are carcinogenic and extremely corrosive. They are fortunately visible, and this should be used as a cue to avoid exposure. Even the equilibrium vapor pressure at 10°C instantly carbonizes organic material; for instance, the inside of polyethylene tubing, while black spots appear on glassware containing even minute traces of organic matter. Thus, it acts as a "fingerprint developer."

REFERENCES

1. Cerfontain, H. and Bakker, B. H., Sulfur trioxide. In *Encyclopedia of Reagents for Organic Synthesis*, edited by Paquette, L. New York: John Wiley & Sons, 2004.
2. Sandler, S. R. and Karo, W., *Organic Functional Group Preparations*, pp. 420–22. New York–London: Academic Press, 1968.
3. Greenwood, N. N. and Earnshaw, A., *Chemistry of the Elements, 2nd ed.*, pp. 703–7. Oxford, U.K.: Butterworth-Heinemann, 1997.

4. Mellor, J. W., *A Comprehensive Treatise on Inorganic and Theoretical Chemistry, Vol. 10*, p. 339. London-New York: Longmans Green, 1930; Tretyakov, Yu. D., Ed., *Praktikum po Neorganicheskoi Khimii* (in Russian), pp. 235–6. Moscow: Academia, 2004.
5. Muller, T. L., Sulfuric acid and sulfur trioxide. In *Kirk-Othmer Encyclopedia of Chemical Technology, 5th ed.*, Vol. 23. New York: John Wiley & Sons, 2006.

21 Thionyl Chloride and Chlorosulfonic Acid

SUMMARY

- Sulfur dichloride reacts with pure SO_3 to yield 62% $SOCl_2$, and pyrosulfuryl chloride.
- Substituting liquid oleum for SO_3 yields the more valuable chlorosulfonic acid instead of pyrosulfuryl chloride.
- The combined yield of $SOCl_2$ and HSO_3Cl with respect to *free* SO_3 is 112% and 78% with respect to total SO_3 in 65% oleum.
- The products are distributed as 63% thionyl chloride, 30% chlorosulfonic acid, 7% pyrosulfuryl chloride.
- $SOCl_2$ and HSO_3Cl/S_2O_5Cl are 99% pure by IC.
- Purification of $SOCl_2$ by double distillation from quinoline gives 53% yield due to ~20% decomposition by distillation from high bp mixtures.
- A laboratory preparation of SCl_2 with a quantitative yield is presented.

APPLICATIONS [1]

Thionyl Chloride
- Substitution of hydroxide group by chlorine, yielding only gaseous by-products.
- Dehydration of ionic compounds, for example, hydrolizable salts of type $MX_n.mH_2O$ [2].
- Dehydration of amides to nitriles.

Chlorosulfonic Acid [3]
- Sulfonation and chlorosulfonation of nucleophiles, such as alcohols and amines as well as activated hydrogen compounds. Acid is more selective than SO_3 and does not attack aliphatic hydrocarbons [4].

Sulfur Chloride [5]
- Synthesis of acid chlorides and anhydrides from alcohols and acids.
- Reaction with aromatics to form mono- and disulfides.
- Reaction with olefins to yield sulfide addition products.

21.1 INTRODUCTION

Thionyl chloride is the mixed-acid anhydride of sulfurous and hydrochloric acids. It is strongly electrophilic and reacts with weak Brönsted acids to yield the corresponding chlorides:

$$ROH + SOCl_2 \rightarrow RCl + SO_2 + HCl, \tag{21.1}$$

$$RCOOH + SOCl_2 \rightarrow RCOCl + SO_2 + HCl. \tag{21.2}$$

The equilibrium is favorably shifted by gas evolution, and unlike PCl_5 and $POCl_3$, there are no nonvolatile by-products. Thionyl chloride reacts vigorously with water ($R = H$ in Reaction 21.1), although often with a considerable delay; hence, the transportation of thionyl chloride is regulated, and it is a sea-freight item.

Chlorosulfonic acid is the mixed acid anhydride of sulfuric and hydrochloric acids used in the chlorosulfonation of organic derivatives. It reacts very vigorously with water to yield HCl and H_2SO_4, as well as with most metals at slightly elevated temperatures, and hence it is a sea-freight item. Sulfur dichloride SCl_2 is a corrosive pungent liquid that hydrolyzes at a moderate rate in water to HCl, SO_2, and colloidal sulfur. Its laboratory preparation is described in the experimental section in a procedure adapted from Schlesinger [5], resulting in a quantitative yield.

The standard procedure for preparing thionyl chloride in the laboratory is chlorination of SO_2 with PCl_5 [6]. This is quite expensive and slow due to the low bp of SO_2, which volatilizes a substantial amount of $SOCl_2$, reducing the yield. An alternative laboratory procedure, also forming the basis of an industrial process, is the co-proportionation of sulfur dichloride with sulfur trioxide [7]:

$$SCl_2 + SO_3 \rightarrow SOCl_2 + SO_2, \tag{21.3}$$

where SO_3 is distilled into the reaction mixture from oleum. There are substantial side reactions, and Michaelis [7] mentions the production of pyrosulfuryl chloride by the action of SO_3 on SCl_4. However, due to the instability of the former, it is more likely that the reaction proceeds by the direct oxidation of $SOCl_2$ with excess SO_3 [6]:

$$SOCl_2 + 2SO_3 \rightarrow S_2O_5Cl_2 + SO_2. \tag{21.4}$$

This reaction, together with a substantial amount of unreacted SCl_2, leads to a maximum yield of 80% with respect to SO_3 reported by Michaelis [7]. Purification from SCl_2 (bp 59°C) is effected by conversion to the high-boiling S_2Cl_2 and distillation, during which the S_2Cl_2 partially disproportionates back to SCl_2.

Here, the procedure of Michaelis [7] with pure SO_3 has been found to yield $SOCl_2$ at ~60–62% on that basis; however, substitution of liquid oleum 65–70% for SO_3 not only simplifies the reaction but also yields the more valuable chlorosulfonic acid instead of pyrosulfuryl chloride as the main by-product. The yield of $SOCl_2$ remains at 62% and the reaction has the following stoichiometry:

$$SCl_2 + 3SO_3 + H_2O \rightarrow 2HSO_3Cl + 2SO_2. \tag{21.5}$$

The chlorosulfonic acid (bp 152°C) is contaminated with ~20% of its anhydride (pyrosulfuryl chloride bp 151°C), which is not significant for most applications, and the two are easily separated from thionyl chloride (bp 77°C) by their widely different boiling points. The overall yield for Reaction 21.3 and Reaction 21.5 is 112% with respect to free SO_3 and 78% with respect to total SO_3 in 65% oleum, with the products distributed as 63% thionyl chloride, 30% chlorosulfonic acid, 7% pyrosulfuryl chloride. With the stoichiometry used in the present experiment, the thionyl chloride contains traces of SO_3 but no substantial amount of SCl_2, as evidenced by lack of free chlorine and the traces of precipitated sulfur at the end of the reaction (which otherwise reacts with SCl_2).

Thionyl chloride can be purified by double distillation from quinoline and linseed oil [8], resulting in a clear, acid-free product; however, the purification is quite wasteful (53% yield for overall procedure) because decomposition of $SOCl_2$, which is often described as slight at near the boiling point [4], has been found to be on the order of 20% for moderate distillation (1 drop/sec.) from the higher-bp mixtures used in Reference 8.

21.2 DISCUSSION

One of the major uses of thionyl chloride is to effect the substitution of the hydroxyl group by chlorine [1]. The reaction proceeds via an intermediate ester:

$$ROH + SOCl_2 \rightarrow ROSOCl + HCl, \qquad (21.6)$$

with subsequent elimination of SO_2, the ester being unstable. However, for certain classes of R, notably aromatics, the ester can be isolated. Thionyl chloride should not be used to substitute tertiary alcohols because its fairly strong dehydrating capacity leads to water elimination; in this case, however, HCl can be used.

A desirable property of Reaction 21.1 is the simple workup afforded when $SOCl_2$ is used in excess. Thus, while high-boiling acid chlorides can be separated from excess $SOCl_2$ by simple distillation, those with a boiling point close to that of $SOCl_2$ (bp 77°C) can be purified by destroying residual $SOCl_2$ using a variation of Reaction 21.2:

$$HCOOH + SOCl_2 \rightarrow CO + SO_2 + 2HCl. \qquad (21.7)$$

Thionyl chloride can be used to form anhydrides directly from organic acid salts without isolating acid chlorides as intermediates [1]:

$$2RCOONa + SOCl_2 \rightarrow RCO\text{-}O\text{-}COR + SO_2 + 2NaCl. \qquad (21.8)$$

This reaction is heavily exothermic and proceeds more readily than methods based on the reaction of an acid chloride with an anhydrous acid salt.

Applications of chlorosulfonic acid include the sulfonation and chlorosulfonation of active hydrogen compounds [3], which are useful intermediates in the synthesis of many organic compounds [4]. Chlorosulfonation requires harsher conditions, with the chlorosulfonic acid used in excess and often at raised temperatures, while sulfonation is carried out generally with a large excess of substrate in a suitable solvent [9]:

$$ArH + 2HSO_3Cl \rightarrow ArSO_2Cl + H_2SO_4 + HCl, \qquad (21.9)$$

$$ArH + HSO_3Cl \rightarrow ArSO_3H + HCl. \tag{21.10}$$

Sulfur monochloride reacts with acids to yield acid chlorides:

$$RCOOH + S_2Cl_2 \rightarrow RCOCl + \tfrac{1}{2}SO_2 + 3/2S + HCl \tag{21.11}$$

and anhydrides:

$$2RCOOH + S_2Cl_2 \rightarrow 2RCO\text{-}O\text{-}COR + \tfrac{1}{2}SO2 + 3/2S + 2HCl.$$

Reaction with aromatics yields mono- and disulfides:

$$S_2Cl_2 + Ph\text{-}H \rightarrow Ph\text{-}S\text{-}S\text{-}Ph + 2HCl. \tag{21.12}$$

With phenols, sulfur monochloride gives condensation products; for example, reaction with C_6H_5OH yields symmetric hydroxyphenyldisulfides [10]. Reaction with olefins yields sulfide addition products:

$$2CH_3\text{-}CH=CH_2 + S_2Cl_2 \rightarrow CH_3\text{-}CH(Cl)\text{-}CH_2\text{-}S\text{-}CH(Cl)\text{-}CH_3 + S. \tag{21.13}$$

21.2.1 THE S/CL$_2$ SYSTEM

Sulfur generally does not react with chlorine in the cold, although the literature contains accounts of reactions commencing this way [11]. The reason for this is that the start of the reaction is an unstable process. If sulfur is present in a particularly reactive form, then a spot reaction generates sufficient local heat to fan the process, as the reaction is exothermic:

$$2S + Cl_2 \rightarrow S_2Cl_2 \qquad \Delta H = -60 \text{ kJ/mol.} \tag{21.14}$$

One does not have to rely on this, and in the present experiment sulfur is heated to 160°C–180°C, generating sufficient vapor pressure to ensure complete absorption of Cl_2 fed into a 1-L flask at the rate of 600 mL/min.

Despite the exothermicity of the reaction, S_2Cl_2 is quite an unstable compound, which decomposes into its elements at 300°C. SCl_2 is more unstable and decomposes slowly even at room temperature (half-life of a few days). In addition, both S and Cl_2 are soluble in S_2Cl_2 to a great extent, corresponding to stoichiometries of about S_5Cl_2 at one extreme, and S_2Cl_5 at the other. Therefore, it is difficult to delineate precisely the amount of mixture and chemical compound in the S–Cl system. Clearly, a substantial amount of the sulfur and chlorine are chemically bound owing to the different physical properties of the product from the reagents, and the sufficiently stable boiling point of S_2Cl_2 (139°C). For many purposes, however, such a delineation is not important. In the present case, as the reaction of SCl_2 with SO_3 reduces the SCl_2 concentration, the dissolved S and Cl_2 react by Le Chatelier's principle, forming more of the reagent.

To enable separation of S_2Cl_2 in the present synthesis, the reaction temperature is maintained at 180°C–210°C, ensuring evaporation of the S_2Cl_2 as soon as it formed.

This arrangement has the additional benefit that impurities, such as ash present in the sulfur, are eliminated.

21.2.2   PURIFICATION OF $SOCl_2$

In addition to the desired product, the following by-products can be present at the end of the reaction: S_2Cl_2, SCl_2, SO_2Cl_2, SO_2, SO_3, S, Cl_2, HSO_3Cl, and $S_2O_5Cl_2$. Further complications arise because $SOCl_2$ decomposes during the distillation:

$$2SOCl_2 \rightarrow SO_2 + Cl_2 + SCl_2. \tag{21.15}$$

The gases SO_2 and Cl_2 present purification difficulties as they are soluble in the liquid products (Cl_2 greatly so; SO_2 must also be soluble, as substantial frothing is not evident in the low-temperature part of the reaction). Similar to all gases, however, they become much less soluble at the boiling point; indeed, SO_2 does not become evident until the reagents are refluxed. SO_3 (bp 44°C) and S_2Cl_2 (bp 138°C) have boiling points sufficiently different from $SOCl_2$ that they are largely removed in the initial distillation. On the other hand, both SO_2Cl_2 (bp 77°C) and SCl_2 (bp 59°C) have boiling points very close to that of $SOCl_2$, but they are unstable and thus are substantially removed by extended refluxing at the end of the reaction. It is sometimes suggested [7,11] that SCl_2 be converted to the higher-boiling S_2Cl_2 prior to fractional distillation; however, the SCl_2 re-forms in the lower half of the distillation column [12]:

$$S_2Cl_2 \rightarrow SCl_2 + S, \tag{21.16}$$

so after distillation, the product is contaminated with an equilibrium mixture of S_2Cl_2–SCl_2.

A colorless product almost completely devoid of the acids SO_3, HSO_3Cl, and $S_2O_5Cl_2$, and the sulfur chlorides is obtained using the purification method suggested in Reference 8. The acids form an adduct with quinoline, which is separated from $SOCl_2$ by distillation. The procedure is quite wasteful (even with half the suggested ratio of quinoline to $SOCl_2$), due to the raised boiling point required to remove the last portion of $SOCl_2$ from the SO_3-quinoline adduct (a solid with white needle-like crystals). The reaction is quite energetic, and loss of thionyl chloride to decomposition, measured as weight loss from the receiver–reactor system, was found to be 15%. A second distillation from boiled linseed oil, used to trap residual sulfur chlorides, is accompanied by some charring and produces a further 28% loss of product, again due to thionyl chloride decomposition during distillation.

21.2.3   SEPARATION OF HSO_3Cl AND $S_2O_5Cl_2$

These compounds are useful by-products of the present preparation, and have almost identical boiling points (151°C–152°C). They are separated as a second fraction in the form of a clear liquid boiling at 145°C–151°C, sp gr 1.75, with the range in the boiling point being due to the mixture. While it is difficult to isolate chlorosulfonic acid, the anhydride can be isolated using its very low hydrolysis rate at subzero temperatures [11]. Thus, hydrolysis with ice in a mixture chilled to −10°C is accompanied by vigorous boiling due to the reaction of the chlorosulfonic acid, while a viscous, slightly yellow lower liquid, separates out, amounting consistently to 20% of the upper fraction weight. The gaseous hydrolysis products were passed into an ice-filled trap, combined with the liquid products and analyzed by ionic chromatography (see Appendix, Figure A.22). The amount of Cl^- and SO_4^{2-} present are in accordance with a HSO_3Cl stoichiometry to within 1%. The high-boiling residue in the reaction flask was found to contain ~10% oleum with a small amount of HSO_3Cl/S_2O_5Cl, which was not distilled to prevent SO_3 contamination of the product (bp 10% oleum 175°C).

21.3   EXPERIMENTAL

21.3.1   PREPARATION OF SULFUR MONOCHLORIDE S_2Cl_2

First, 98.5 g of 99% flowers of sulfur is put in a two-neck, 250-mL flask placed on a heating mantle. One neck is equipped with a gas inlet tube reaching within a few millimeters of the bottom, and fed with chlorine gas dried by means of a single H_2SO_4 scrubber with a glass wool spray trap. The other neck is connected through a water-cooled Liebig condenser to a 250-mL receiver, whose outlet is protected from moisture by a P_2O_5 trap.

The heating mantle is set to about 40% duty cycle (on a 280 W mantle), as moisture is removed from the apparatus by means of a slow nitrogen flux through the chlorine generator head. Once the sulfur has melted, the nitrogen is replaced by a fairly rapid chlorine flow (about 600 mL/min, or 2 drops/sec. of 15% HCl onto 232 g TCCA in the chlorine generator). After an initial lag, while the S_2Cl_2 product dissolves in the sulfur, drops of a yellow-orange fluid start to gather in the receiver at about 0.5 drops/sec. (see Figure 21.1). The flask temperature continues to rise to about 200°C due to the heat of the reaction, and beyond 160°C there is no longer any chlorine color at the reactor outlet. Testing with saturated NaCl solution confirms that no Cl_2 passes the flask unreacted.

After about 400 mL of 15% HCl has been added (90 min), the distillation flask is completely empty, save for a few particles of soot. The product in the receiver flask changes color from yellow to orange in the last stages of the reaction, as free chlorine passing the flask reacts with the S_2Cl_2 to form a small amount of SCl_2, whose red streaks slowly diffuse from the surface. This has the benefit of removing all traces of sulfur from the apparatus. The yield of sulfur chloride is 206.4 g, or 99% S_2Cl_2 with respect to sulfur.

21.3.2   PREPARATION OF SULFUR DICHLORIDE SCl_2

Begin by placing 205.5 g of S_2Cl_2 product from the previous preparation into a 1-L, two-neck flask cooled by an ice water reservoir and fitted with a gas inlet tube reaching to its bottom (Figure 21.2). The other flask neck is connected to a 300-mm reflux, double-wall condenser cooled by ice water from the reservoir, and protected from

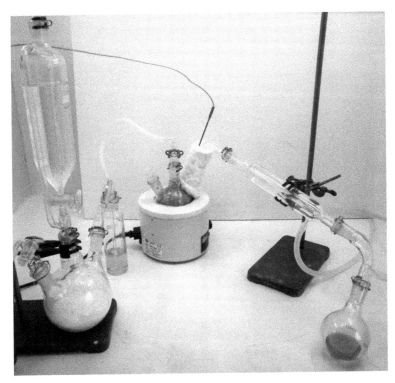

FIGURE 21.1 *A color version of this figure follows page 112.* Chlorine gas reacts instantly with sulfur at 160°C, producing S_2Cl_2 if sulfur is present in excess. The orange liquid S_2Cl_2 collects in the receiver; however, its color starts to deepen toward the end of the reaction as excess chlorine oxidizes the sulfur further to the deep red SCl_2.

moisture by a P_2O_5 trap. The low temperature ensures maximum rate of chlorine dissolution in the sulfur chloride and minimum decomposition of the SCl_2 product, while the flask provides a large volume for a vapor phase reaction as well as a large surface area for absorption of gaseous chlorine by liquid condensed on its walls.

A small amount (0.1 g) of steel wool is now added to the flask [11], substantially enhancing absorption of the chlorine. At the end of the experiment, the steel wool is only superficially reacted; hence, the catalysis is due to trace amounts of $FeCl_3$ and no purification by distillation is required. Chlorine is now introduced at 600 mL/min (as in the previous section) and is completely absorbed with no chlorine gas evident in the reflux condenser (color), while very few chlorine bubbles break through the surface of the sulfur chloride. After approximately 400 mL of HCl has been consumed in the chlorine generator, substantial reflux commences in the reagent flask due to consumption of S_2Cl_2 and a corresponding drop in the boiling point of the mixture, with chlorine gas starting to appear at the outlet of the condenser. Generation of chlorine is now stopped and the weight of the deep red liquid product is found to be 324.5 g, corresponding to a 100% yield of SCl_2 with 3.4% dissolved chlorine (clearly this figure varies with the exact point at which the reaction is stopped). The

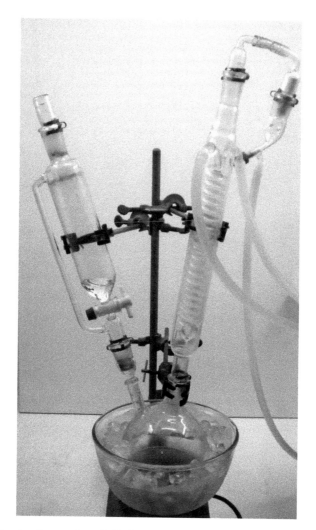

FIGURE 21.2   *A color version of this figure follows page 112.* Thionyl chloride and chlorosulfonic acid are the major products when cooled sulfur dichloride is oxidized by sulfur trioxide, which is introduced here in the form of liquid oleum. Large amounts of sulfur dioxide are also produced, initially dissolving in the reaction mixture. The equilibrium is shifted to the right by heating. All tubing is polyethylene, as silicone is rapidly degraded by sulfur trioxide vapor.

dissolved chlorine can be regarded as SCl_4, although the pure compound is unstable above $-30°C$ [8].

21.3.3   Preparation of $SOCl_2$ and HSO_3Cl

The gas inlet tube is now replaced by a dropping funnel fitted with a Teflon cock (this avoids charring of grease associated with a glass cock) and charged with 237 g of 65% oleum (1.93 mol free SO_3). Then, 224.5 g (2.18 mol) SCl_2 is kept in the flask, and a 3-cm Teflon stirrer is introduced. As the flask is cooled in an ice water mixture, oleum is dripped into the SCl_2 at a rate of about 1 drop/sec with rapid stirring. Some solidification occurs as mentioned by Michaelis [7,11], but as this occurs entirely in a thin layer on the flask walls, it is likely to be unreacted oleum rather than any solid product. A small amount of blue-colored liquid may appear in SO_3 drops deposited on the flask walls, due to the formation disulfur trioxide by the reaction of a small amount of sulfur present in SCl_2 (or generated by the decomposition of S_2Cl_2) with SO_3.

After all the oleum has been added (about 30 min), the temperature of the water bath is raised to 75°C in the course of about 40 min. At about 40°C, the flask contents begin to froth as SO_2, which carries with it some SO_3, starts to evolve, driving the equilibrium in Reaction 21.5. No chlorine evolution should be observed at any stage during the procedure. After a final temperature of 75°C is reached, gentle reflux (1–2 drops/sec.) is continued for about 30 min to remove as much excess SCl_2 and SO_2Cl_2 as possible from the mixture (no further SO_3 is evolved beyond about 50°C). The reaction should not be continued further, since although SO_2 evolution does not diminish, at this stage this is due to the decomposition of $SOCl_2$ by reflux rather than formation of the product by Reaction 21.5. The flask contents weigh 328 g at this stage, corresponding to a weight loss of 133.5 g due mainly to evolved SO_2 and excess SCl_2, as well as $SOCl_2$ lost by decomposition.

The apparatus is now arranged for fractional distillation through an ice water-cooled Graham condenser, and discarding the first few drops of SO_3 and SCl_2 in the lowest fraction, two fractions boiling between 70°C–80°C and 145°C–152°C are collected. The low-boiling fraction yields 136.1 g, corresponding to crude $SOCl_2$ (1.14 mol), containing a small amount of SO_3, SO_2Cl_2, and SCl_2/S_2Cl_2 impurity, while the high-boiling fraction yields 78.5 g, corresponding to a 80:20 mixture of HSO_3Cl (0.54 mol) and $S_2O_5Cl_2$ (0.07 mol). 45 g of ~10% oleum (bp 175°C) remains in the flask, with 68 g lost to gaseous products in the fractional distillation.

The combined yield of $SOCl_2/HSO_3Cl/S_2O_5Cl_2$ with respect to free SO_3, via the Reaction 21.3–Reaction 21.5, is therefore 112%. This figure is explained by the fact that the residue of 10% oleum contains 31 g less SO_3 than in the H_2SO_4 content of the original oleum. This signifies that more than the free SO_3 content of oleum is active in the present reaction, and explains the high yield with respect to free SO_3. It is nevertheless appropriate to quote the yield on this basis because alternative preparations of the foregoing products operate on a free SO_3 basis.

The gaseous weight loss in the course of the reaction amounts to 201.9 g. Of this, 118 g is accounted for by SO_2 released in proportion to the products collected from Reaction 21.3–Reaction 21.5. A calculation shows that the remaining weight loss is made up of volatilized 71 g SCl_2 and 13 g SO_3 either free or in the form of the products of Reaction 21.3.

21.3.4   PURIFICATION OF SOCL$_2$

As mentioned in the introduction, thionyl chloride can be purified by distilling first from quinoline, then from boiled linseed oil [8]. In the present procedure, the ratio of the high-boiling compound to $SOCl_2$ has been decreased somewhat; nevertheless, the entire process gives a yield of only 53%.

To 135.5 g $SOCl_2$ in a 250-mL, two-neck flask connected to a 250-mL receiver by means of an ice water-cooled Graham condenser, and protected from the atmosphere by means of a P_2O_5 plug, is added dropwise with stirring, 13 g of quinoline. The reaction generates considerable heat, which causes some of the $SOCl_2$ to boil. When all the quinoline has been added, the flask and associated adaptors are thermally insulated and the $SOCl_2$ is distilled slowly, boiling at 75°C–77°C (about 8% duty cycle on a 280 W heating mantle) until no more product comes over in this temperature range. At the end, the heating needs to be raised to about 25%, and some of the adduct sublimes inside the flask. Then, 108.8 g of product is collected, representing a yield of 80%, while the distillation flask loses 116.7 g, so that about 6% of the thionyl chloride is lost by decomposition.

The distillation flask is now replaced by the receiver flask, to which 25 g of boiled linseed oil is added, and the thionyl chloride distilled as before. Then, 71.6 g of colorless liquid boiling in the range 77°C–78°C is collected, representing a yield of 67% for this step, and a yield of 53% for the purification procedure overall. The weight loss of the distillation flask is 99.6 g, so that about 26% thionyl chloride is lost to decomposition in this step.

S_2Cl_2 and SCl_2, similar to all sulfides, are poisonous, with a pervasive odor resembling sewer gas. They dissolve plastics and latex—a stray drop on a glove generates palpable heat at that spot within seconds. Crude $SOCl_2$ is a yellow, mobile, fuming liquid that has a strong odor of SO_2 and sulfur chlorides (as is the commercial technical product). It rapidly penetrates greased quickfit joints and its fumes attack metals readily. Due to the presence of a small quantity of acids, it readily chars many organic compounds. The purified product is colorless with an odor of HCl and SO_2.

REFERENCES

1. Wirth, D. D., Thionyl chloride. In *Encyclopedia of Reagents for Organic Synthesis*, edited by Paquette, L. New York: John Wiley & Sons, 2004.
2. Greenwood, N. N. and Earnshaw, A., *Chemistry of the Elements, 2nd ed.*, p. 694. Oxford, U.K.: Butterworth-Heinemann, 1997.
3. Prabhakaran, P. C., Chlorosulfonic acid. In *Encyclopedia of Reagents for Organic Synthesis*, edited by Paquette, L. New York: John Wiley & Sons, 2004.
4. McDonald, C. E., Chlorosulfuric Acid. In *Kirk-Othmer Encyclopedia of Chemical Technology, 5th ed.*, Vol. 6. New York: John Wiley & Sons, 2006.
5. Schlesinger, G. G., *Inorganic Laboratory Preparations*, pp. 119–21. New York: Chemical Pub. Co., 1962.
6. Weil, E. D., Sandler, S. R., and Gernon, M.., Sulfur Compounds. In *Kirk-Othmer Encyclopedia of Chemical Technology, 5th ed.*, Vol. 23. New York: John Wiley & Sons, 2006.
7. Michaelis, A., Über die Thionylamine. *Eur. J. Org. Chem. (Ann. Chem.)* 274(2): 173–86, 1893.
8. Furniss, B. S., Hannaford, A. J., Smith, P. W. G., and Tatchell, A. R., *Vogel's Textbook of Practical Organic Chemistry, 5th ed.*, p. 466. London: Addison Wesley Longman, 1989.

9. Cremlyn, R. J., *Chlorosulfonic Acid: A Versatile Reagent*, pp. 35–60. Cambridge, U.K.: The Royal Society of Chemistry, 2002.
10. Masami, H., A Method for the Preparation of a Bis(hydroxyphenyl) Sulfide. EU Patent No. EP0287292, Apr. 13, 1987.
11. Fehér, F., Sulfur, selenium, tellurium. In *Handbook of Preparative Inorganic Chemistry, 2nd ed.*, pp. 382–3, p. 387, edited by Brauer, G. New York–London: Academic Press, 1963.
12. Kunkel, K. E., Purification of Thionyl Chloride. U.S. Patent No. 3155457, Nov. 3, 1964.

Appendix: Assay of Reagents

A.1 INTRODUCTION

Since the present preparations are intended as alternatives to reagent purchase, their assay is an essential part of this book. Because an in-house preparation would not generally be considered for applications requiring extra-high purity reagents (better than about 99.9%), the aim of the present chapter is to provide an assay of synthesized reagents to a level of about 0.1%.

Three types of analysis suitably cover the types of reagents prepared here. The majority of reagents, such as the alkali metals of Chapters 2–5, are rapidly and completely hydrolyzed by water, yielding ions whose assay can be used to determine the composition of the reagent. Indeed, many of these chemicals are so reactive that any other type of analysis would be exceedingly complicated. Basic reagents are assayed for cationic species, while ionic chromatography is carried out for acidic reagents. The several types of salts prepared are assayed for both anions and cations. The ionic chromatograph used here is associated with sensitivities in the range of 10–100 ppb for most ionic species of interest, and a dynamic range of $>10^5$, substantially exceeding that of ICP machines.

A second type of reagent, such as carbon disulfide and carbon tetrachloride of Chapters 14 and 16 is nonionic and does not react with water. These reagents are volatile and, therefore, suitable for analysis by GC/MS, enabling a large range of organic content to be investigated. A third type of reagent is represented by triethylaluminum of Chapter 10 and triphosgene of Chapter 7, which are nonionic but is also unsuitable for GC/MS (triphosgene decomposes in the injection port). The nonintrusive methods of proton and ^{13}CNMR were used to analyze these reagents for both organic and hydrogen content. HNMR is also a particularly sensitive assay for carbon disulfide, carbon tetrachloride, and triphosgene, which do not contain hydrogen in the pure state, enabling the extent of chlorination to be measured accurately in the latter two cases, and the amount of thiophene and mercaptans to be established in the former case.

More details on the methods used to perform the analyses, their basis of operation, the preparation of reagents, and the standards employed are given under the specific headings in the following text. These are followed by figures of the spectra of the various reagents with explanations given in the captions.

A.2 SUPPRESSED IONIC CHROMATOGRAPHY WITH CONDUCTIVITY DETECTOR

The Dionex ICS-1600 suppressed ionic chromatography system used here with AS12A and AS12C 4 mm × 200 mm columns relies on high-pressure eluent flow

(~1 mL/min) through a 9 μm macroporous resin functionalized with quaternary ammonium groups for anions and carboxylic/phosphonic acid groups for cations. The eluent consists of an 8 mM:1 mM $Na_2CO_3/NaHCO_3$ buffer for the anionic column and 20 mM CH_4SO_4 for the cationic column, enabling the detection of ionic concentration largely independent of the pH and counter-ion content of the original sample. Detection is by means of a temperature-compensated conductivity cell.

Since the eluent is itself highly conductive, system sensitivity can be greatly improved if this conductivity is canceled prior to the flow entering the detector. This can be achieved by neutralizing the eluent in an electrolyzing suppressor, thus generating CO_2 from the anionic and H_2 from the cationic eluents. Both of these compounds outgas, producing a solution of much lower conductivity. In the anionic case, this technique also enhances the sample signal because, with the CO_3^{2-} ion replaced by sample anions (e.g., Cl^-), the introduction of hydrogen ions does not decrease conductivity by releasing CO_2, but instead increases it. In the cationic case, the hydrogen ions should be the most electronegative cations present; hence, suppressed detection is not suitable for cations, such as Cu^{2+}, which are reduced preferentially to hydrogen. Transition metals, such as Fe^{2+}, are also unsuitable for suppressed detection as they precipitate hydroxides in the suppressor. Nonetheless, the suppressed conductivity system used here has been found sensitive to electropositive transition metals, such as Fe^{2+} and Zn^{2+}, at the ppm level. High concentrations of these ions should, however, be avoided.

The sensitivity of the system has been found to be about 30 ppb for sodium ions (level of system noise) and, hence, the instrument has a very large useful dynamic range of about 10^5. The sensitivity decreases even in molar terms for larger cations due to decreased mobility. For cations such as $N_2H_5^+$, the sensitivity is reduced by a factor of ~40 with respect to sodium, due mostly to its small dissociation constant and low reduction potential.

The separating capability of the column is dependent on the ionic species. The AS12C is optimized for alkali, alkali earth metals, ammonium, and quaternary ammonium ions. Different species of transition metal ions do not separate well. A column, such as the CS5, can separate transition metals using oxalic acid complexing, etc.; however, since they do not function well with suppressed conductivity detection, AA was used for individual ions of interest, such as Cu^{2+}.

A.2.1 SAMPLE PREPARATION AND MEASUREMENT

Hydrolysis was carried out in deionized water with subsequent dilution with conductivity water to instrument levels of ~100 ppm for the major ions. Because this reaction is extremely vigorous for substances, such as cesium and chlorosulfonic acid, water was added gradually and wide-bore ice-packed glass tubing was used to trap spray. If this is not done, substantial variation exists between samples. The solution resulting from the hydrolysis of alkali metals was acidified prior to testing for transition metals to ensure solvation of these ions.

The response of the conductivity cell is monotonic but nonlinear in the concentration range of interest, and there exists some interference between ions, especially when the concentration of one ion greatly exceeds another. For this reason, measurements were made using dual standards, with the first standard used to evaluate approximate ion concentrations and a second standard made to approximate the sample with respect to all species present.

A.3   HNMR AND ^{13}CNMR

Measurements were made on the Bruker Avance 400 MHz Ultrashield spectrometer using TopSpin software. Signal shimming was performed automatically in the Z-direction only, while baseline and phase adjustments as well as peak integration were done manually. Because the NMR is being applied quantitatively, the main aim of the adjustments was to achieve spectral resolution of the proton signals of interest, and then determine whether any reasonable variation of parameters changes the integral ratio of the various peaks. Peak separation for the present relatively simple spectra was easily achieved, and proton NMR provided stable peak ratios accurate to within a few percent between experiments. Due to the well-known large and variable relaxation times of carbon atoms in various environments, peak ratios in ^{13}C spectra were not found to converge better than ~20% even at a 10 sec. delay between samples. These spectra, therefore, were used qualitatively.

Due to the extreme reactivity of triethylaluminum/diethylbromoaluminum and n-butyllithium, these samples were run without an internal frequency-lock standard. With the high temporal frequency stability of the Bruker instrument, it was found that preliminary signal locking onto a CDCl$_3$ standard followed by immediate measurement of the sample produced results indistinguishable in accuracy from those obtained using an internal standard.

The HNMR signal was found linear to within a few percent with respect to hydrogen concentration for signals up to about 10 mol% hydrogen content. Moreover, with an experiment duration of 1 min (16 samples), the instrument has a very large dynamic range of about 10^4 (down to about 0.001 mol% hydrogen); hence, all measurements were made with respect to a single internal concentration standard.

A.4   GAS CHROMATOGRAPH—MASS SPECTRA GC/MS

Measurements were made on a Varian CP-3800 gas chromatograph coupled to a Saturn 2200 mass spectrometer with CP-8400 auto-sampler. Samples were run at injection temperatures of 208°C (lower temperatures did not achieve complete volatilization for some samples evidenced by a drop in flow rate) with a linear oven temperature variation of 72°C/min, after an initial hold of 2 min at 56°C. The diluent was chloroform with the sample diluted to about 1% and the chloroform signal blanked over a 0.1 sec. interval. The rather low dilution ratio was required to enable resolution of impurities present at the 0.1% level (e.g., C_2Cl_6 in CCl_4). Electronic ionization was used, and all significant spectral peaks were positively identified by spectrometer software. The ratio of peak integrals obtained from the mass spectrometer particle detector corresponds only to within an order of magnitude to specimen molar ratios for the types of molecules tested, and standards were required to achieve more accurate determination.

A.5 REAGENT SPECTRA AND ASSAY

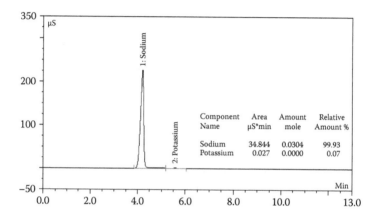

FIGURE A.1 Cation chromatograph of sodium for hydrolysis products (Chapter 2). Cu^{2+} < 0.1% by AA.

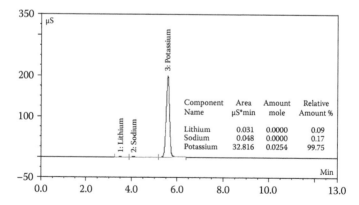

FIGURE A.2 Cation chromatograph of potassium hydrolysis products (Chapter 3). Cu^{2+} < 0.1% by AA.

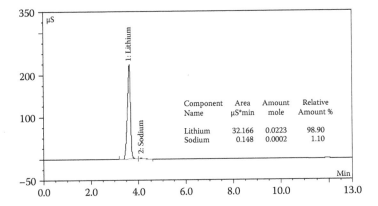

FIGURE A.3 Cation chromatograph of lithium hydrolysis products (Chapter 4). $Cu^{2+} < 0.1\%$ by AA.

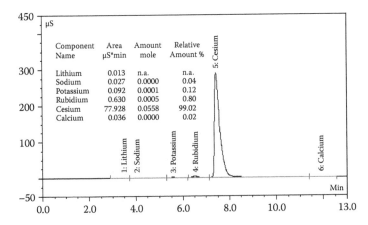

FIGURE A.4 Cation chromatograph of cesium hydrolysis products (Chapter 5). $Cu^{2+} < 0.1\%$ by AA.

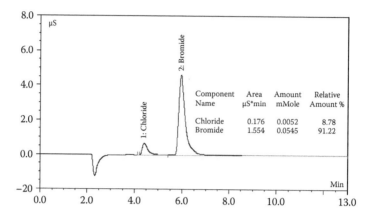

FIGURE A.5   Cation chromatograph of triethylaluminum hydrolysis products (Chapter 10), based on 0.136 g (1.2 mM). The assay suggests the presence of ~4.5% diethylaluminum bromide.

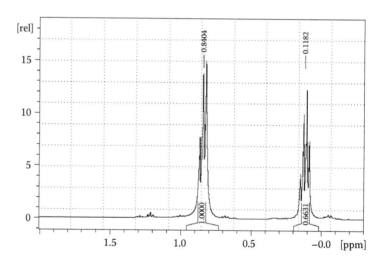

FIGURE A.6   HNMR spectrum of diethylaluminum bromide neat (Chapter 10) showing a reasonably pure compound with some peak broadening due to intermolecular interaction. The CH$_3$ triplet is at 0.84 ppm and the CH$_2$ quadruplet at 0.12 ppm. The integrals are in the expected 2:3 proportion.

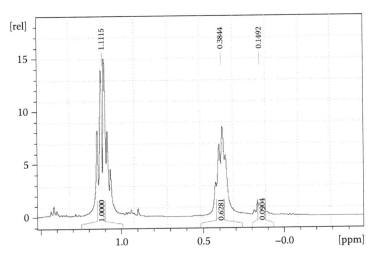

FIGURE A.7 HNMR spectrum of triethylaluminum neat (Chapter 10) with ~4.5% Et₂BrAl impurity (see Figure A.5). The CH$_3$ group signal of Et$_2$BrAl is convoluted with that Et$_3$Al at ~1.11 ppm, manifest by the substantial splitting of the Et$_3$Al CH$_3$ group and only a residual signal left at 0.84 ppm. The substantial interaction makes the spectrum of qualitative value only.

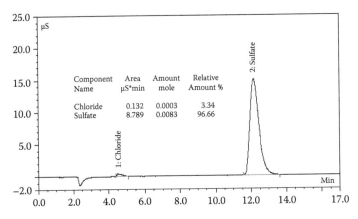

FIGURE A.8 Anion chromatograph of hydrazinium sulfate solution (Chapter 11), based on 1.09 g (0.00838 mol). The assay shows the presence of 3 mol% Cl⁻ ion (3 mol% NaCl based on cation chromatograph).

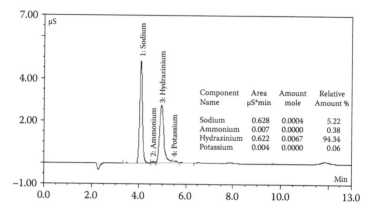

FIGURE A.9 Cation chromatograph of hydrazinium sulfate solution (Chapter 11) based on 0.98 g (0.0071 mol). The assay shows the presence of 5 mol% Na⁺ ion.

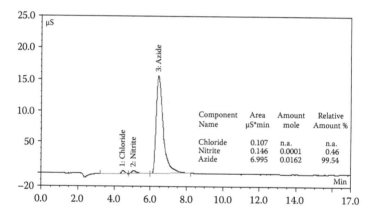

FIGURE A.10 Anion chromatograph of potassium azide solution (Chapter 12) based on 1.284 g (0.0158 mole). Combined with cation chromatograph, Figure A.9, overall purity is > 99%. The ~2.5% difference in chromatograph and gravimetric concentrations is due to experimental error.

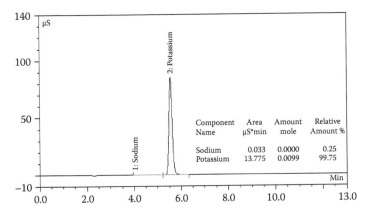

FIGURE A.11 Cation chromatograph of potassium azide solution (Chapter 12) based on 0.826 g (0.010 mole).

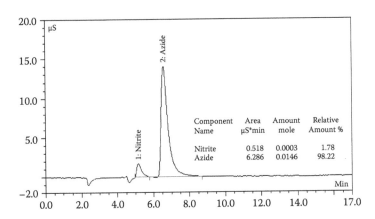

FIGURE A.12 Anion chromatograph of sodium azide solution (Chapter 12) based on 0.956 g (0.0146 mole). The assay shows ~1 mol% nitrite impurity.

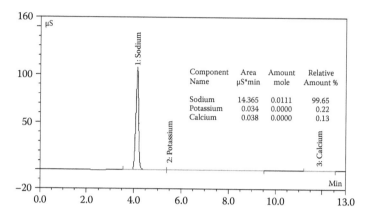

FIGURE A.13 Cation chromatograph of sodium azide solution (Chapter 12) based on 0.734 g (0.011 mole).

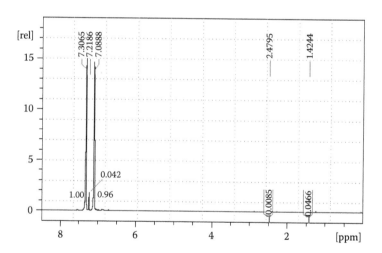

FIGURE A.14 HNMR spectrum of 0.337 g carbon disulfide (Chapter 14) in 1.066 g CDCl$_3$ reference containing 0.13% CHCl$_3$. The 7.09 and 7.31 ppm signals are from the two pairs of equivalent protons in thiophene C$_4$H$_4$S, while the 7.22 ppm signal is from chloroform. A ratio of peak integrals gives the thiophene content at ~3.5%.

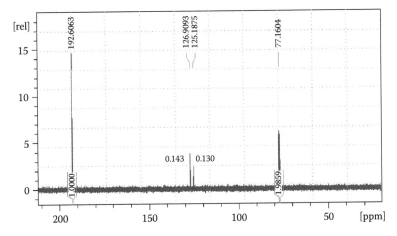

FIGURE A.15   ¹³CNMR spectrum of carbon disulfide (Chapter 14). The 192 ppm signal is from CS₂, while the 125 ppm and 127 ppm signals are from thiophene, C₄H₄S. With 77 ppm being the CDCl₃ reference, all other organics are constrained below ~ 0.5%. Peak integrals given are not a quantitative indication because the ¹³CS₂ signal is diminished by its slow relaxation.

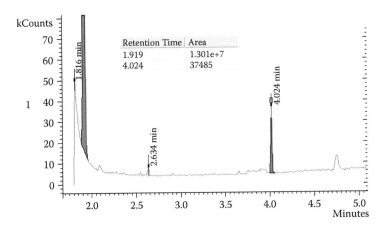

FIGURE A.16   GC/MS spectrum of carbon tetrachloride (Chapter 16) solved in chloroform. The peak at 1.91 sec. corresponds to CCl₄, at 2.63 sec. to CH₃COCl from the eluent, and at 4.02 sec. to C₂Cl₆. The latter concentration is determined at 0.1% by reference to a standard. Chloroform is the only other significant impurity and is determined by HNMR in Figure A.17.

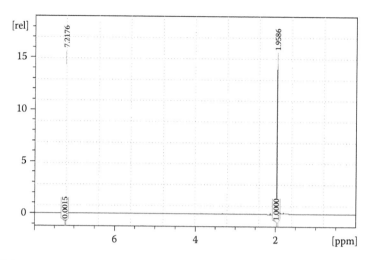

FIGURE A.17 HNMR spectrum of 1.159 g carbon tetrachloride (Chapter 16) with 0.0154 g acetone. Acetone provides a useful single-peak standard at 1.96 ppm, while 7.21 ppm is the signal from residual chloroform. A ratio of peak integrals gives the chloroform concentration as 0.1%.

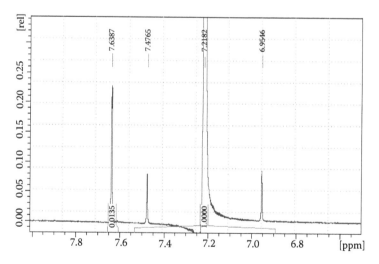

FIGURE A.18 HNMR spectrum of 0.135 g triphosgene (Chapter 17) dissolved in CCl_4 with 0.102 g $CHCl_3$ internal standard. The signal at 7.22 ppm corresponds to $CHCl_3$ (6.95 ppm and 7.48 ppm are the ^{13}C splitting), while the signal at 7.64 ppm shows the presence of a single, significant, nonexhaustively chlorinated intermediate, $C_3HCl_5O_3$. A ration of peak integrals gives the latter concentration at 2.2%. No other hydrocarbon impurities are present.

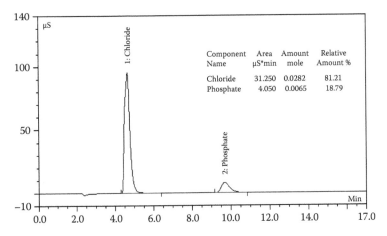

FIGURE A.19 Anion chromatograph for the hydrolysis products of 1.216 g (0.0058 m) PCl₅ (Chapter 18). The small difference in the measured PO_4^{3-}: Cl⁻ molar ration from the expected 1:5, and the excess phosphate shown in the chromatograph, correspond to the presence of up to 10% PCl₃, generated when PCl₅ is sublimed. The former hydrolizes to phosphite, which elutes together with phosphate and/or oxidized to the latter.

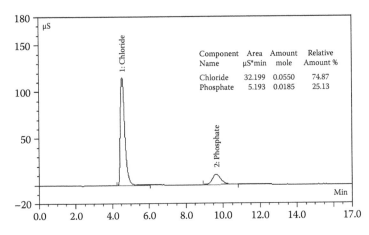

FIGURE A.20 Anion chromatograph of POCl₃ hydrolysis products (Chapter 19). The phosphate to chloride ratio is the expected 1:3 within experimental error, no other peaks are present. This assay does not distinguish between POCl₃ and PCl₃; however, the latter is constrained by redox titration.

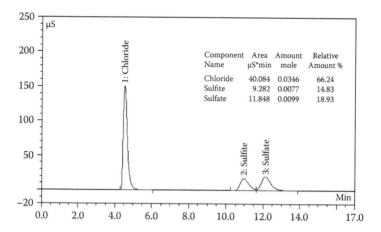

FIGURE A.21 Anion chromatograph of $SOCl_2$ hydrolysis products (Chapter 21). Dual sulfite/sulfate peaks are due to about half the sulfate formed initially by hydrolysis being oxidixed to sulfate during shaking in the presence of atmospheric oxygen while the sample is prepared (samples are only 100 ppm, and the sulfate content decreases with the amount of shaking). The overall sulfate + sulfite to chlorine ratio is the expected 1:2 within experimental error. This assay constrains SO_3, HSO_3Cl, S_2O_5Cl, and HCl impurities; it does not constrain SO_2Cl_2.

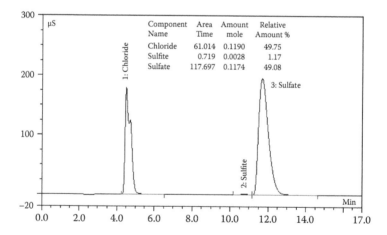

FIGURE A.22 Anion chromatograph of HSO_3Cl hydrolysis products (Chapter 21). The SO_4^{2-}: Cl^- ratio is the expected 1:1 within experimental error, and about 1–2% $SOCl_2$ impurity is indicated.

Index

Milton Keynes UK
Ingram Content Group UK Ltd.
UKHW051015071024
449327UK00012B/257